Researching Pedagogy and Practice with Canadian Mathematics Teachers

A Volume in:
Research in Mathematics Education

Series Editors
Denisse R Thompson
Mary Ann Huntley
Christine Suurtamm

Research in Mathematics Education

Series Editors

Denisse R Thompson
University of South Florida

Mary Ann Huntley
Cornell University

Christine Suurtamm
University of Ottawa

International Perspectives on Mathematics Curriculum (2018)
Denisse R Thompson, Mary Ann Huntley, & Christine Suurtamm

Digital Curricula in School Mathematics (2016)
Meg Bates & Zalman Usiskin

Approaches to Studying the Enacted Mathematics Curriculum (2012)
Dan Heck, Kathryn Chval, Iris Weiss, & Steven W. Ziebarth

Variability is the Rule:
A Companion Analysis of K-8 State Mathematics Standards (2010)
John P. Smith III

Language and Mathematics Education:
Multiple Perspectives and Directions for Research (2010)
Judit N. Moschkovich

A Five-Year Study of the First Edition of the Core-Plus Mathematics Curriculum (2010)
Harold Schoen, Steven W. Ziebarth, Christian R. Hirsch, & Allison BrckaLorenz

Future Curricular Trends in School Algebra And Geometry:
Proceedings of A Conference (2010)
Zalman Usiskin, Kathleen Andersen, & Nicole Zotto

A Decade of Middle School Mathematics Curriculum Implementation:
Lessons Learned from the Show-Me Project (2008)
Margaret R. Meyer & Cynthia W. Langrall

Mathematics Curriculum in Pacific Rim Countries—China, Japan, Korea, and Singapore:
Proceedings of a Conference (2008)
Zalman Usiskin & Edwin Willmore

The History of the Geometry Curriculum in the United States (2008)
Nathalie Sinclair

The Classification of Quadrilaterals: A Study in Definition (2008)
Zalman Usiskin

The Intended Mathematics Curriculum as Represented in State-Level Curriculum Standards
Consensus or Confusion? (2006)
Barbara Reys

Researching Pedagogy and Practice with Canadian Mathematics Teachers

Edited by

David A. Reid
Christine Suurtamm
Annie Savard
Elaine Simmt
Dominic Manuel
Lisa Lunney Borden
Richard Barwell

IAP
INFORMATION AGE PUBLISHING, INC.
Charlotte, NC • www.infoagepub.com

Library of Congress Cataloging-In-Publication Data

The CIP data for this book can be found on the Library of Congress website (loc.gov).

Paperback: 978-1-64802-137-4
Hardcover: 978-1-64802-138-1
E-Book: 978-1-64802-139-8

Printed and bound by CPI Group (UK) Ltd, Croydon, CR0 4YY

CONTENTS

PART II

PART III

ACKNOWLEDGMENTS

The research reported in this book was supported by the Social Sciences and Humanities Research Council of Canada.

Special thanks to Susan Mielke for her work in copyediting the manuscripts for this book.

PREFACE

David A. Reid

This book is based on the Observing Teachers research program, a collaborative research project that examined middle school mathematics pedagogy in four regions of Canada[1]. Using an innovative methodology adapted from anthropology, we explore the assumptions, beliefs, and values that affect teachers' actions, often without self-awareness of their influence. We refer to these assumptions, beliefs, and values as *pedagogy*. A fundamental challenge facing us as researchers is that pedagogy cannot be directly observed or reported introspectively. In this book, we offer results from the project related to the participating teachers' pedagogy, comparisons grounded in an assumption that regional communities of practice and pedagogy exist in Canada, observations on the methodological issues we faced, and explorations of themes triggered by the research. This book will be of interest to researchers in mathematics education, teacher educators, teachers, and government policy makers.

Each chapter of this book was written by members of the research team, but no chapter would have been possible without the participations of the full team. Here and henceforth all members of the team should be considered to have contributed to all aspects of this book.

Chapter 1 outlines the motivation for the project with respect to the differences in achievement among Canadian provinces. It includes some background infor-

[1] For more information, see http://www.math.uni-bremen.de/~dreid/OT/. This research is supported by the Social Sciences and Humanities Research Council of Canada (SSHRC grant #410-2011-1074).

Researching Pedagogy and Practice with Canadian Mathematics Teachers, pages ix–xi.

mation about Canada that is important to understanding the context of the project. Next, the theoretical background, which is grounded in the anthropological work of Tobin (1989), Tobin, Hsueh, and Karasawa (2009), and Tobin, Wu, and Davidson (1989) is described, followed by methodological considerations, which are grounded in the work of Maturana (1987, 1988, 2002), Maturana and Varela (1992), and Varela (1999). The research design and data collection methods used in the project conclude the chapter.

The research design partly guides the structure of the book. Teachers met in focus groups, first to discuss videos of lessons taught by teachers in the group. These discussions we refer to as *auto-ethnographies*. The teachers met again to discuss videos of lessons taught by teachers in other groups. These discussions we refer to as *ethno-ethnographies*. These discussions were analyzed both by researchers who took part in collecting the data, whom we describe as having an *insider* perspective on the data, and by other members of the research team, whom we describe as having an *outsider* perspective on the data.

Part I contains chapters in which insiders analyze the ethnographies. In Chapter 2, Elaine Simmt explores the Alberta teachers' notions of pedagogy as inferred from comments they made when viewing videotapes of their own and other teachers' lessons. In Chapter 3, Christine Suurtamm and Brenna Quigley describe the pedagogy that four Ontario teachers revealed in their viewing of their own lessons. In Chapter 4, members of the McGill research team, Annie Savard, Dominic Manuel, Sophie Pinard, Scosha Merovitz, and Vandana Chandrasekhar, analyze the discussions of teachers in three focus groups through an analytic framework blending elements from French *didactique* (e.g., Brousseau, 2006) and Lampert et al.'s (2013) ambitious pedagogy. They use this analysis to further develop their framework. In Chapter 5, Sophie Pinard, Dominic Manuel, and Annie Savard focus on three groups' discussions of teaching and learning fractions. This concentration on the discussions of lessons with related content allows for a focused analysis.

Part II contains three chapters in which outsiders analyze the ethnographies. The chapters all examine the same data, auto-ethnographies from English speaking teachers. In Chapter 6, Richard Barwell and Lisa Lunney Borden take a discursive approach focusing on teachers' constructions of common pedagogy in their discussions of videos of each other's teaching. In Chapter 7, Lisa Lunney Borden and Ellen Carter examine the data thematically. In Chapter 8, David Reid takes a topical and visual approach to identify similarities and differences in the data. This part opens and closes with brief notes on connections among the chapters.

Part III includes reflections on the research in two chapters. In Chapter 9, two teachers who participated in the project reflect on how it changed the way they see teaching, with framing comments provided by some members of the McGill research team. In Chapter 10, Richard Barwell weaves together reflections from the full research team to reflect on and theorize some of the many challenges that arose during the Observing Teachers project.

Chapter 11 concludes the volume. In this chapter, David Reid reflects on all the chapters and the research project as a whole. He (re)considers the methodology of the project given its outcomes, how those outcomes address the original question guiding the research, as well as its grounding assumptions and definitions.

REFERENCES

Brousseau, G. (2006). *Theory of didactical situations in mathematics: Didactique des mathématiques, 1970–1990* (V. Warfield, R. Sutherland, M. Cooper, N. Balacheff, Trans.). Dordrecht, the Netherlands: Springer.

Lampert, M., Franke, M. L., Kazemi, E., Ghousseini, H., Turrou, A. C., Beasley, H., ... Crowe, K. (2013). Keeping it complex: Using rehearsals to support novice teacher learning of ambitious teaching. *Journal of Teacher Education, 64*(3), 226–243.

Maturana, H. (1987). Everything is said by an observer. In W. I. Thompson (Ed.), *Gaia: A way of knowing: Political implications of the new biology* (pp. 65–82). Great Barrington, MA: Lindisfarne Press.

Maturana, H. R. (1988). Reality: The search for objectivity or the quest for a compelling argument. *The Irish Journal of Psychology*, *9*(1), 25–82.

Maturana, H. R. (2002). Autopoiesis, structural coupling and cognition: A history of these and other notions in the biology of cognition. *Cybernetics & Human Knowing*, *9*(3–4), 5–34.

Maturana, H. R., & Varela, F. J. (1992). *The tree of knowledge: The biological roots of human understanding* (Rev. Ed.). Boston, MA: Shambhala.

Tobin, J. (1989). Visual anthropology and multivocal ethnography: A dialogical approach to Japanese preschool class size. *Dialectical Anthropology*, *13*(2), 173–187.

Tobin, J., Hsueh, Y., & Karasawa, M. (2009). *Preschool in three cultures revisited: China, Japan, and the United States.* Chicago, IL: University of Chicago Press.

Tobin, J. J., Wu, D. Y. H., & Davidson, D. H. (1989). *Preschool in three cultures: Japan, China, and the United States.* New Haven, CT: Yale University Press.

Varela, F. J. (1999). *Ethical know-how: Action, wisdom, and cognition.* Stanford, CA: Stanford University Press.

CHAPTER 1

OBSERVING TEACHERS

Comparing Middle School Teachers' Pedagogies in Four Regions of Canada

David A. Reid

The Observing Teachers research program was a collaborative research project that examined middle school mathematics pedagogy in four regions of Canada. This introduction sets out some of the background context and shared research foci and methods that defined the research program at the outset. The research grew and evolved as it progressed, and the chapters of this book provide insights into that evolution.

BACKGROUND

Large-scale international and national assessments have revealed a considerably wide range of student achievement in mathematics across Canada. When compared to international results, some Canadian provinces, notably Quebec rank among the top countries, while other provinces, especially in the Atlantic region, fall significantly below the Canadian average. There are also some differences related to the language of schooling. Students who are educated in French in Quebec and in English in Ontario achieve a higher average than their peers schooled in the other language in the same province (Brochu, Deussing, Houme & Chuy, 2013). These differences have been evident in international assessments since

Researching Pedagogy and Practice with Canadian Mathematics Teachers, pages 1–21.

2000 (Bussière et al., 2001) and in national assessments over a longer period (e.g., Council of Ministers of Education, 1997).These differences have been attributed to a number of factors, including student gender, attitudes, beliefs, aspirations and time spent working outside school; parents' education, involvement in their children's education, and socio-economic status; and school curriculum and resources (see Anderson et al., 2006; Beaton & O'Dwyer, 2002; Schmidt et al., 2001; Wilkins, Zembylas, & Travers, 2002). Teaching, which might be expected to have the most direct effect on student achievement, is less often considered. This led us to consider ways in which we might conduct a cross-Canada comparative study of the teaching of mathematics.

Our Definition of Pedagogy

The teaching of mathematics, that is, the observable behaviors of teachers, varies considerably from teacher to teacher. Furthermore, the same teacher may teach different topics, or different groups of students, in distinctly different ways. This makes directly comparing teaching practice between regional or linguistic groups quite difficult. However, we assume that underlying a teacher's behaviors is a set of beliefs, tacit knowledge, customs and habits that guide practice while remaining largely implicit, and that teachers who interact regularly will evolve pedagogies that are similar. This provides our initial definition of *pedagogy*:

> **Pedagogy** refers to the set of (largely implicit) beliefs, knowledge, customs and habits that guide a teacher's practice, and also the shared implicit beliefs, knowledge, customs and habits of a community of teachers who interact regularly.

This concept of pedagogy resembles part of what Tobin, Hsueh, and Karasawa (2009) call the "'implicit cultural practices' of teachers, ... practices that though not taught explicitly in schools of education or written down in textbooks reflect an implicit cultural logic" (p. 19). These practices are not limited to teachers' behaviors, but include both their actions and intentions. As Tobin et al. (2009) note, these implicit practices are related to teachers' "knowledge in practice" (Anderson-Levitt, 2002, p. 109) and "embodied knowledge" (Anderson-Levitt, 2002, p. 8). Such knowledge is related to Bruner's (1996) concept of folk pedagogy, the "taken-for-granted practices that emerge from embedded cultural beliefs about how children learn and how teachers should 'teach'" (p. 46). In the following, we use the term *pedagogy* to refer to such implicit elements that influence teaching.

Tobin et al. (2009) define *practice* as "action plus intention" (p. 19). We prefer to distinguish action and intention, and to use *behavior* and *pedagogy* to refer to these two aspects of what Tobin et al. call practice. Pedagogy could also be described as the *professional culture* of mathematics teachers; however, the phrase professional culture is used in another sense in the education literature (e.g., Lieberman, 1988).

We further theorize pedagogy as being an *emotional orientation* to particular practices. Maturana (1988) uses his concept of emotional orientation to explain science as a domain of explanation, that is, as a domain in which characteristic implicit criteria for making and accepting explanations apply. The same idea applies to teaching as a domain of practice. Maturana argues that the criteria for accepting an explanation in science must be implicit at some level, in order to escape an infinite regress. Explicit criteria at some level require an explanation that must be based on more fundamental criteria. Those more fundamental criteria, if they are explicit, necessitate the introduction of even more fundamental criteria. The only way to escape an infinite regress of more and more fundamental criteria is to have implicit criteria at some level that remain unexplained and hence do not require more fundamental criteria for their explanation. Similarly, the criteria for teachers' in-the-moment decision making about practice must also be implicit at some level. Explicit principles are both impractical to access in the ongoing flow of teaching (Brown & Reid, 2004, 2006) and would be a part of the teacher's explicit practice and so require deeper criteria. Later we will discuss how we use teachers' watching of videos to reveal the implicit criteria underlying their teaching.

Membership in a community is established on the basis of shared emotional orientations. This is a recursive process that evolves over time, so that individuals adjust their emotional orientations to fit the communities they find themselves in and, at the same time, their emotional orientations contribute to the community's standards of behavior. Just as populations of animals become more distinct and eventually evolve into new species when isolated from contact with others, this recursive process of evolving emotional orientations means that communities become more distinct when isolated from contact with other communities. This is the theoretical basis for our assumption that in different regions of Canada different pedagogies will occur.

This brings us to a second definition of pedagogy founded on Maturana's work:

> **Pedagogy** refers to the emotional orientation towards teaching that guides a teacher's practice, and also to the shared emotional orientation towards teaching of a community of teachers who interact regularly.

The word pedagogy in English is used infrequently, often in the sense of the theoretical basis of the practice of teaching which is not far from our usage. Hence, it seems to us a suitable label for the concept we wish to study. However, in French, *pédagogie* is often contrasted with *didactiques*, with *pédagogie* referring to teaching methods and theory independent of subject matter, and *didactiques* referring to teaching methods and theory specific to a subject. This presents a problem for researchers in our group who wish to make this distinction in their analysis (see Chapter 4). We are also not the first researchers to assign a technical meaning to pedagogy. Van Manen (2008), for example, uses the word in his work in a sense derived from its etymology, as "guiding children" (Van Manen, 2008, p. 6). This use of the word has also come into our analyses, especially in Chapter 2.

Regional Variation in Canada

We assume that variations in pedagogy occur among the regions of Canada while not ignoring that differences within regions also occur. Hence, it is useful to consider aspects of the cultural geography of Canada that could give rise to such variations. As the cultural geography of Canada is tied to its physical geography, a few aspects of this physical geography are outlined first.

Physical Geography, Climatic, and Linguistic Variations Across Canada

Canada is the second largest country on Earth in area but much less populated than similarly sized areas like Europe, China, and the United States. The low population density of Canada is both caused and mitigated by Canada's climate, which is cold. Most Canadians live in the southern, warmer regions, which spreads most of the population out over what is effectively a long thin country 200 km wide and 5000 km long. This southern region is interrupted by three physical features: the Rocky Mountains in the west, the Canadian Shield in the center, and the Gulf of Saint Lawrence in the east. The Canadian Shield is a region of exposed rock with a thin layer of soil making it unsuitable for agriculture. Driving from Sudbury, Ontario to Winnipeg, Manitoba across this region, a distance of about 1700 km, one passes through only two cities of any size, both on the shores of Lake Superior. These physical features contribute to the separation of regions of Canada.

Compounding the physical features that separate regions of Canada, Canada has two official languages, French and English, that are not evenly distributed across the country. The province of Quebec is officially francophone (French speaking) and 80% of the population speaks French at home. New Brunswick, east of Quebec, is officially bilingual and 32% francophone. The other provinces are either officially anglophone (English speaking) or have no official language and have small francophone minorities. The right of students who speak the minority language at home to be educated in their own language is constitutionally guaranteed, and so there are schools operating in both languages in all provinces.

These physical, climactic, and linguistic circumstances divide Canada into four regions: the mostly anglophone provinces on the Atlantic coast (from which the province of Newfoundland and Labrador is partly separated by the Gulf of Saint Lawrence), francophone Quebec, anglophone Ontario, and (separated from Ontario by the Canadian Shield) the anglophone provinces in western Canada (from which the province of British Columbia is partly separated by the Rocky Mountains). Travel between these regions is difficult. Toronto, the largest city in Ontario, is 2700 km away from Calgary (Alberta), the largest city in western Canada, that is, Ontario is farther from Calgary than London is from Moscow.

Political Geography, School Organization, and Teacher Preparation in Canada

A further factor giving rise to regions is the political geography. Each of Canada's ten provinces sets its own standards for education, including curriculum,

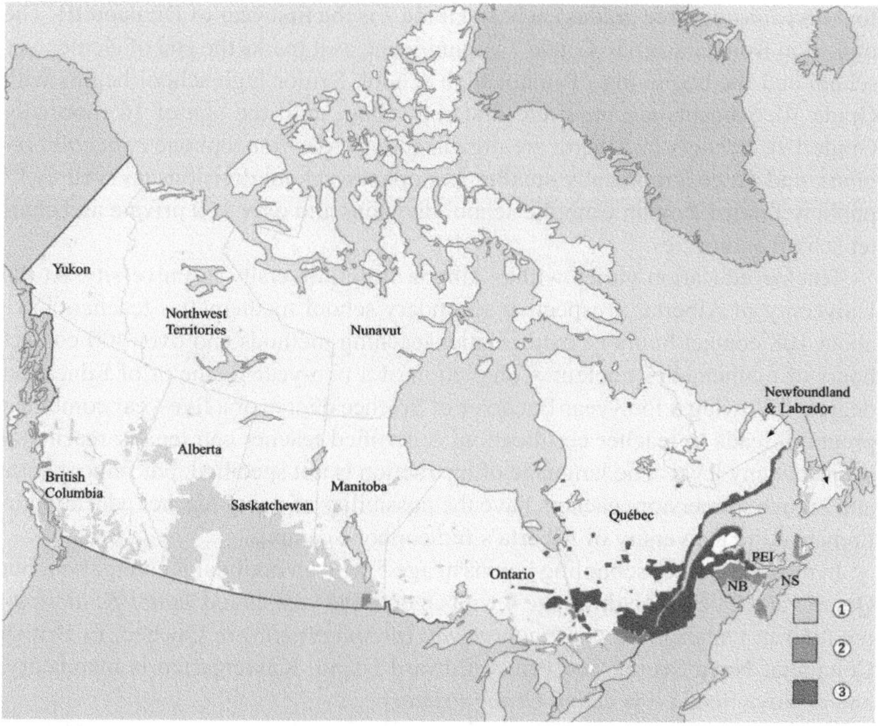

FIGURE 1.1. Primary language spoken in different regions of Canada. English (1), Mixed (2), French (3). The unshaded region has a population density of less than 0.4 per km^2 and the primary languages spoken and the languages of instruction include English, French, Cree, Inuktitut, and other Indigenous languages. NB: New Brunswick; PEI: Prince Edward Island; NS: Nova Scotia. Adapted from: https://commons.wikimedia.org/wiki/File:Bilinguisme_au_Canada-fr.svg

school organization, and teacher preparation. Historically, every province had its own mathematics curriculum, but this changed in the 1990s when two regional groupings formed, in the West and in the East, in which provinces developed and implemented mathematics curricula collaboratively. At that time, the curricular map of Canada reflected the four regions described above. Since around 2010, all of the Atlantic provinces have adopted the curriculum used in the western provinces. Simmt (2018) provides a nuanced description of the nature of curriculum in Canada.

One way to see the range of approaches to school organization and teacher preparation across Canada is to begin with a single province and then describe how other provinces differ from it. This is the approach we take here, using Alberta as an example.

In Alberta, schooling begins at about age 5, with an optional single year of Kindergarten. This is followed by twelve grades, each a year long, organized into four *divisions* of three grades each, so Grade 7 is the first year of Division III. The transition from Grade 6 to Grade 7 is significant, as it marks the end of elementary school and the beginning of junior high school. Senior high school begins with Grade 10. Students are required to attend school until the age of 16 (normally Grade 10). Schools in Alberta are organized into four francophone *education regions,* and 42 geographically smaller anglophone school divisions, as well as 17 publicly funded Roman Catholic school divisions and over 160 private and charter school authorities.

Teacher education varies within Alberta from university to university. At the University of Alberta, prospective secondary school mathematics teachers have about 108 contact hours of mathematics teaching methods and over 400 contact hours of mathematics content. Completion of a two-year Bachelor of Education degree, following a four-year Bachelor of Science degree or a five-year combined program, leads to teacher certification. A certified teacher can legally teach any subject at any level. The language of instruction is not specified, but francophone and French immersion teachers have the possibility of doing teacher education in French at the University of Alberta's francophone campus.

In most provinces schooling begins at age 5 with an optional Kindergarten, but Ontario and Quebec begin at age 4 with an optional year called *Junior Kindergarten (JK)* in Ontario and *pre-kindergarten* (*prématernelle*) in Quebec. In British Columbia, Nova Scotia, and Prince Edward Island, Kindergarten is mandatory, and in Nova Scotia it is called *Grade Primary*.

As in Alberta, many provinces have six-year elementary schools (*l'école primaire* in French). In Quebec, these six years are organized into three 2-year *cycles,* so what is Grade 5 in many provinces is the first year of Cycle 3 in Quebec. The end of elementary school occurs earlier than Grade 6 in some provinces. In British Columbia, Grade 4 is the first year of *intermediate*; in Manitoba, Grade 5 is the first year of *middle school*; in Saskatchewan, New Brunswick, and some schools in Nova Scotia, Grade 6 is the first year of middle school. In Ontario, Grades 7–8 are middle school but are sometimes grouped with the elementary grades in a single JK–8 school. In Quebec, Grades 7 and 8 are referred to as Secondary I and II and constitute Secondary Cycle 1.

The next transition from intermediate / middle / junior high / Secondary Cycle 1 to secondary/ senior high / Secondary Cycle 2 occurs at the start of Grade 9 in British Columbia, Manitoba, Ontario, Quebec, and New Brunswick. As in Alberta, students in most provinces are required to attend school until the age of 16, but in Manitoba, Ontario, and New Brunswick schooling is mandatory to age 18. The last year of school is Grade 12 everywhere except Quebec where it ends with Grade 11 (Secondary V, or the third year of Secondary Cycle 2).

Most provinces have fewer school authorities than Alberta. Ontario, the most populous province, has 72 school *boards*[1], 60 English language and 12 French

[1] Corresponding to what is called a school *district* in the US, or a *local education authority* in the UK.

	Alberta	Ontario	Quebec	New Brunswick
First year of public school	Age 5 Kindergarten	Age 4 Junior Kindergarten	Age 4 pre-Kindergarten (*prématernelle*)	Age 5 Kindergarten
First compulsory year of school	Age 6 Grade 1	Age 6 Grade 1	Age 6 Grade 1	Age 6 Grade 1
Groupings of school years	Elementary school (Division I, Grades 1–3; Division II, Grades 4–6) Junior high school (Division III, Grades 7–9) Senior high school (Division IV, Grades 10–12)	Elementary school (Grades 1–6) Middle school (Grades 7–8) Senior high school (Grades 9–12)	Elementary school (*école primaire*, Cycle I, Grades 1–2; Cycle II, Grades 3–4; Cycle III, Grades 5–6) Secondary (Cycle 1, Sec. I–II, Grade 7, 8; Cycle 2, Sec. 2, III–V, Grades 9 –11)	Elementary school (Grades 1–6) Middle school (Grades 7–8) Senior high school (Grades 9–12)
Last compulsory year of school	Age 16 (normally Grade 10)	Age 18 (normally Grade 12)	Age 16 (normally Grade 10)	Age 16 (normally Grade 10)
Last year of public school	Grade 12	Grade 12	Sec. V (Grade 11)	Grade 12
Jurisdictions responsible for schools	francophone education regions, anglophone school divisions, Roman Catholic school divisions, private and charter school authorities	English language and French language school boards	francophone *commissions scolaires,* anglophone school boards, Cree and Inuit school boards	anglophone and francophone school districts

FIGURE 1.2. School organization in Canada (for provinces represented in the Observing Teachers project)

language. In Quebec, there are 60 francophone *commissions scolaires* and nine anglophone school boards, as well as two school boards in the north of the province serving the Cree and Inuit Nations. The smaller provinces have proportionally fewer school authorities. In New Brunswick, there are four anglophone and three francophone school districts; in Nova Scotia, a single French language school board covers the entire province and English language schools are organized into seven regional districts.

Teacher preparation differs across Canada, though these differences have decreased over time. Reid (2004) reports a range for future elementary school teachers from about 20 contact hours in mathematics teaching methods (and no mathematics content) at some universities to 200 hours of methods and content at others. The variation is likely less at the secondary level, and provincial certification requirements require more hours than at elementary and are roughly the same across Canada. However, provinces differ in how subject specific their teaching certification is. In most provinces, as in Alberta, a certified teacher may teach any subject at any grade level, and in situations where there is a shortage of qualified mathematics teachers, middle school mathematics might be taught by a teacher whose education was focused on elementary school teaching, or on teaching science or another subject. In Quebec, however, teaching certification specifies the subjects, grades, and language of instruction the teacher is certified to teach in.

Summary

The organization of education at the provincial level means that the officially constituted communities for education are provincial. Mathematics teachers are most likely to interact with other teachers within their schools and these interactions may influence their pedagogies in ways that are independent of the subjects they teach. Their interaction with other mathematics teachers is likely to occur at school district, provincial, or regional professional development sessions, and mathematics specific conferences. School districts in Canada are usually language specific, and so such interaction is likely to be with teachers who share the same language. Interaction with teachers from other provinces, especially from provinces in other regions of Canada, is unlikely due to the physical, climactic, linguistic and political circumstances described.

METHODOLOGY

The guiding question for this research is: How do pedagogies in middle school mathematics in regions of Canada differ? As noted previously, pedagogies are implicit, and this presents a challenge for research. How do we observe what is not explicit?

Our research methodology evolved from an epistemological position summarized in the observation that "everything is said by an observer" (Maturana, 1987, p. 65). Maturana's work in the neuroscience of perception led him to conclude that he "had to abandon the question, 'how do I see that color?' and ask instead, 'what happens in me when I say that I see such a color?'" (Maturana, 2002, p. 5). Instead of asking questions about an external world, the focus shifts to asking questions about observers and the worlds they observe.

What does it mean to research pedagogy from this perspective? It means that it makes little sense to go into classrooms or to view video recordings of classroom teaching and attempt to reconstruct the pedagogy underlying the teaching observed. Such observations are productions of the observer and not a reflection of the teaching observed. However, it seems likely that a teacher's pedagogy would not only guide her teaching but also guide her observations of teaching. This suggested the notion of putting teachers in the role of observers of their own or other teachers' teaching practice and making their observations the basis for our research. Soon after this idea occurred to us, we encountered Tobin and colleagues' work (Tobin, 1989; Tobin, Hsueh, & Karasawa, 2009; Tobin, Wu, & Davidson, 1989) which starts out from a different theoretical position but pursues a similar approach.

To investigate implicit practices, Tobin, Hsueh, and Karasawa (2009) focus not on behaviors, but on teachers' comments on their own and other teachers' behaviors recorded on video.

> In our analyses, we give greatest emphasis not to what teachers in each culture do, but to how they think about what they are doing. Our focus is primarily not on teachers' behaviors, but on their practice or *praxis*, which we define as action plus intention. The videotapes capture behaviors; it is only when we add the explanations of the teachers and directors that we can get at practice. (p. 19)

For instance, for many mathematics teachers, a typical behavior might be to present only correct answers on the chalkboard. When they are confronted with a video showing a teacher intentionally leaving an incorrect student-generated answer on the board, their discussion of the video reveals their own implicit cultural practices.

Tobin (1989) was dissatisfied with both the traditions of positivist anthropology and reflexive approaches. Positivist "cross-cultural research follows the form of Westerners studying non-Westerners, whites studying non-whites, and again, scholars studying practioners [*sic*] and men writing about (or just as often, failing adequately to write about) the lives of other cultures' women and children" (Tobin, 1989, p. 173). The power inequities in this model are clearly problematic and were Tobin's main concern. From our epistemological position, positivist research methods are also inadequate because all that is achieved is observers reporting their self-generated observations and pretending they are reflections of the worlds they observe.

According to Tobin (1989):

> Reflexivity generally involves a focus either on the limitations of the individual researcher including ethnocentrism, professional anxiety and ambition, and counter-transference reactions or on limitations of the discipline of anthropology including conservative, self-perpetuating conventions of discourse, intellectual elitism, and a tendency towards colonialism, sexism, and orientalism. … we are bothered by the irony that reflexivity, while successfully undercutting the researcher's omnipotence, often also has the effect of centering the text even more dramatically on the figure of the researcher-author. (pp. 173–174)

In other words, reflexivity acknowledges that everything said is said by an observer but reacts by making the observer the object of the research, thus neglecting the original object of the research.

Tobin (1989) refers to his method as *populist post-structuralism* or *multivocal ethnography*. Key principles are "a belief in the ability of ordinary people to explain the meanings of their actions and to analyze the institutions of which they are a part" (p. 174) and "a rashomonian[2] telling and retelling of the same event from different perspectives; an ongoing dialogue between insiders and outsiders, between practioners [*sic*] and researchers, and between people of different cultures" (p. 176).

RESEARCH DESIGN

Tobin (1989) describes his research approach in this way:

> Our approach begins with the making of videotapes in comparable settings in two or more cultures. For example, for our current research, we shot videotapes of typical days in preschools in Japan, China, and the United States. Next, we take edited versions of these tapes back to the settings where they were filmed and ask insiders to give their interpretations and explanations. For the third step in our method, we take videotapes shot in one culture to the other cultures in the study, asking, for example, Japanese and American preschool teachers to make judgments about a tape we made in a Chinese preschool. (p. 176)

Tobin (1989) calls the step of making videotapes *visual ethnography*, asking insiders to give their interpretations he calls *auto-ethnography*, and asking outsiders to give their interpretations he calls *ethno-ethnography*. We will adopt these terms to describe our own methods.

In designing our research, we followed the structure outlined by Tobin and colleagues (Tobin, 1989; Tobin, Wu, & Davidson, 1989). We made video recordings of middle school teachers in the four regions of Canada previously described, teaching in three contexts: a lesson the teacher would describe as typical of his/her teaching; a lesson the teacher considered exemplary in some way; and a lesson on the topic of fractions. These contexts were chosen in order to generate a variety

---

2 This word refers to the 1950 Kurosawa film Rashomon, in which several characters provide distinct and contradictory versions of the same incident.

of prompts to be used in the later stages and also as an early indication of what the teachers felt was typical and exemplary teaching. The meaning of *typical* and *exemplary* was left to the teachers to decide. The lesson on fractions was included to ensure that at least some of the videos would involve similar content so that differences related to the content taught would not distract from other differences. As we knew it would be impossible to have videos in which the teachers taught exactly the same content in all the classrooms, the topic of fractions is sufficiently common in middle school curricula across Canada that the fraction lessons would likely include content all the teachers had taught at some point.

Unlike Tobin, we did not use the visual ethnographies, the videos of classroom teaching, as part of our data. As outlined, this reflects our positioning of the teachers as observers and ourselves as observers of their observations. It also addresses a practical concern, that is, that many school districts and university research ethics boards in Canada are wary of video recording in schools. In fact, it was emphasized in our research design that only a clearly defined group of people, that is the researchers and the teachers in the focus groups, would be given access to the classroom videos.

The following procedure was followed to make the videos of classroom teaching. Each teacher chose the lessons to be recorded, and either the teacher or a research assistant recorded the lessons, with a single camera focused on the teacher. In Ontario, both a researcher and a research assistant were present and they set up two video cameras so that they captured two different angles and had a backup if one recording was faulty. The video was then edited to 20 minutes or less by a research assistant, in consultation with the teacher who selected the segments to be included in each video. The edited videos presented the lessons in an accessible format, which included key episodes and omitted sequences (e.g., individual paper-and-pencil tasks) that are not usefully portrayed visually. Omitted sequences were marked with descriptive titles. The final edited videos were shown to the teacher to check that they portrayed the lessons adequately. The intent was to produce a video short enough to be comfortably viewed in the focus groups, but which also was acceptable to the teacher of the lesson.

In the next stage, corresponding to Tobin's (1989) auto-ethnographies, the teachers in each focus group viewed the edited videos featuring their classrooms and attempted to identify three videos (from among all the videos made in the classrooms of the teachers in the group) to be shared with teachers in other regions. The teachers were instructed to select one video in each of three categories: a typical lesson, an exemplary lesson, and a lesson about fractions. As before, the meaning of *typical* and *exemplary* was left to the teachers to decide. The recordings of these focus group discussions formed our first data set: as responses of observers within each regional and linguistic group, they provided an auto-ethnography of mathematics teaching in each region. No discussion prompts for these sessions were decided in advance, beyond the goal of selecting the videos to be shared. Each researcher guided these sessions differently, according to local circumstances. These differences are discussed below and in subsequent chapters.

We then shared the three videos chosen for each region with the corresponding linguistic groups in all the other regions. In Tobin and colleagues' (Tobin, 1989; Tobin, Wu, & Davidson, 1989) terms, they acted as stimuli for the ethno-ethnography phase. In this stage, we expected encounters with other teachers' teaching to offer the participants a way to reflect on their own familiar beliefs and practices in comparison with those held by others. The recordings of these focus group discussions formed our second data set. They were analyzed to identify points of difference and similarity observed by the participants.

As noted previously, school districts and university research ethics boards are sometimes wary of video recording in schools. This concern is not uniform across Canada, but it did influence our research design. The limits set on the use of the videos and the decision to edit them were both guided by concern over obtaining approval to conduct the research. The editing process not only allowed us to shorten the recordings, thus reducing the time required for the focus groups to view them, but it also meant we could assure research ethics boards, school districts, school administrators, and parents that no students would be recognizable in the edited videos, either because episodes showing students would be edited out or because faces would be blurred. The limiting of the audience for the videos to the researchers and participating teachers was also intended to ensure the privacy of the teachers and students involved. This contrasts strongly with Tobin and colleagues' (Tobin, 1989; Tobin, Wu, & Davidson, 1989) use of video, which was to make his research more accessible to a wide audience. In our case we chose a design that would better our chances to be allowed to actually conduct the research, at the cost of limiting the ways in which we could communicate our results as the videos would not be shared with others in conference presentations or professional development.

One decision we made in our research design was to use videos recorded in the classrooms of the participating teachers instead of adopting Chazan and Herbst's design (Chazan, Herbst, & Sela, 2011; Herbst & Chazan, 2006, 2011). Chazan and Herbst created animations of teaching episodes to use as prompts in teacher professional development. The animations often include sequences that Chazan and Herbst consider breaches of normal practice. They have identified a number of advantages related to using hypothetical animations of classroom interaction rather than actual videos of live classroom activities. They perceive the verisimilitude of live video as being a disadvantage in their work, "in that video can too forcefully narrate one story, address one context, and thus obliterate any need to [relay] alternative stories that should have happened instead or could have happened in another context" (Herbst & Chazan, 2006, p. 216). For our research, however, we do not envisage any disadvantages to using live video. In fact, because we expect alternative stories of what could have happened to come from the teachers' different pedagogies, it is important that the videos offer authentic portrayals of teaching. Moreover, unlike Chazan and Herbst, we do not want to predetermine what practices will provoke reactions. We do not know what teachers in one region will see as significant when watching a video of a teaching activity in another region. We hypothesized that what they do find significant would, in part, reveal their pedagogy. In the next section, we include some comments from the teachers that illustrate this.

SPECIFICS OF DATA COLLECTION

The following focus groups (described in more detail below) were recruited:

- Alberta anglophones (three teachers);
- Ontario anglophones (four teachers);
- Quebec anglophones (three teachers);
- Quebec francophones (four teachers); and
- New Brunswick francophones (four teachers).

The size of the focus groups was determined by the practicalities of making and distributing the videos. We wanted the groups to be large enough to have a variety of viewpoints and teaching styles represented, but the time needed to view the videos made by the teachers in each group (even in their edited forms) meant that groups larger than four were impractical.

We attempted to include focus groups from both language groups in the four regions. We were unable to recruit francophone focus groups in the western provinces and in Ontario. In Ontario, there was strong support from the school district, but not enough francophone teachers volunteered to make a focus group possible. We hypothesize that a reluctance to be video recorded while teaching was a factor in some teachers' decisions not to participate.

In Atlantic Canada, we attempted to form focus groups of teachers in Nova Scotia, the most populous Atlantic province, but a number of changes were being made in the mathematics curriculum at the time we began our research. The school boards refused permission for their teachers to participate, as they feared that our research would distract teachers from implementing the curriculum changes. Based on our previous experience with conducting research using video in Nova Scotia, we also suspect that concerns over classroom video recording may have been a factor. We were able to form a francophone focus group in New Brunswick composed of teachers who were comfortable with being videoed. Hence, the francophone focus group in New Brunswick is the only one from Atlantic Canada.

We had planned that francophone groups would view francophone videos and anglophone groups would view anglophone videos, to ensure good comprehension of the language of instruction. However, in Quebec the francophone teachers themselves asked if it were possible to view the videos from the anglophone Quebec teachers.

Focus Group	Videos Viewed by the Focus Group
Alberta anglophones (ABA)	ONA, QCA
Ontario anglophones (ONA)	ABA, QCA
Quebec anglophones (QCA)	ABA, ONA, QCF, NBF
Quebec francophones (QCF)	QCA, NBF, ONA, ABA
New Brunswick francophones (NBF)	QCF, QCA, ONA, ABA

FIGURE 1.3. Focus Groups and Videos Viewed

Johanne: Même si ce sont des vidéos en Anglais, on aimerait bien savoir ce qu'ils font eux. [Even if the videos are in English, we would like to know what they are doing.]

Carine: Oui tu sais, dans le fond, on a peut-être les mêmes programmes ici au Quebec. Mais on a aucune idée de ce que les enseignants anglophones font. J'aimerais bien connaitre leurs idées à eux aussi. On pourrais-tu voir aussi les vidéos des anglophones, surtout de notre province? Ou devons-nous juste voir ceux des francophones des autres provinces? [Yes, you know, in the end, we may have the same curriculum here in Quebec. But we have no idea what anglophone teachers are doing. I'd like to know their ideas, too. Could we see the videos of anglophones, especially in our province? Or do we just see those of francophones in other provinces?]

The francophone groups in Quebec and New Brunswick and the anglophone group in Quebec felt sufficiently confident in their understanding of the other language to view videos from the other language group. We considered subtitling if it would make viewing videos in the other language easier for the teachers, but they indicated that it would not be necessary. Figure 1.3 shows the focus groups and which videos each group saw from the other groups.

In all cases, the participants were made fully aware of the goals of the research and all assumptions about the study, including its design and definitions of pedagogy, prior to participating in the study. Standard research ethics protocols were followed.

DATA COLLECTION IN EACH REGION

Details of the methods of data collection were left open to accommodate local conditions. As a result, somewhat different procedures for recording and editing the videos were followed in each region. As the videos were used as prompts for the focus group discussions, rather than being data themselves, we do not see this variation as problematic. One example has already been mentioned, that in Ontario two cameras were used. In this section, we describe the specifics of the data collection and the composition of the focus groups.

Alberta Anglophones

The three teachers in this focus group did not know each other before joining the focus group. They were recruited primarily through school principals, who passed information about the study along to teachers. One teacher was doing a master's degree and was recruited directly. The teachers all came from different schools, with a range of socio-economic contexts. We refer to the teachers by the pseudonyms Brenda, Maryanne, and Andrea. More detail about their backgrounds is given in Chapter 2.

The research assistant for this focus group was a Master of Education student and high school teacher in the same school district as those in the study. This person was selected in part because she spoke French (as we had originally intended to recruit a francophone group in Alberta).

We hoped to minimize disruption of the teachers' teaching when recording the videos. However, in one case in Alberta, the process of obtaining permission to video record in classrooms and negotiating times for video recording delayed the recording to the last month of classes before the summer break. The teacher, wanting to participate in the study, agreed to do review lessons to fit the three lesson types we planned to record. In the other Alberta classrooms, the fraction lessons were done to fit the recording schedule, not as part of the normal sequence of teaching. In this sense, at least five of the nine lessons recorded in the three Alberta classrooms could be viewed as staged lessons. Two of the teachers gave a lesson on fractions in spite of it not being part of their yearly plan. The third teacher, however, chose to focus on a problem that required rational number thinking because she felt that a lesson on fractions, "was out of synch with my year plan… and I wasn't at a point I could even review, it would have seemed crazy." One of the other teachers seemed surprised by her colleague's concern as that teacher's students "wanted to keep going with reviewing fractions." The limited number of days for recording lessons also meant that some lessons were atypical in other ways. One teacher commented: "Well, it was a weird day, because a big chunk of them were gone on a band trip and they [the remaining students] were afraid to talk in front of the camera. Normally they are not afraid to talk."

A single video camera was used to record the lessons. This was set up by the research assistant at the back of the room, where images of students would be kept at a minimum. The research assistant edited out parts of the videos that showed the students working individually at their desks before the teachers reviewed the videos. This was done to save the teachers' time but, as it turned out, created some problems. When the teachers were presented with the edited videos they were disturbed by the absence of student desk-work and conversations. As noted above, we asked each focus group to select three videos to share with the other groups, including one that should portray a lesson they all felt was typical. The teachers in Alberta were reluctant to select a single video as typical because they all taught differently. They asked if they could create a combined video that demonstrated the teaching style of all three teachers. The research team agreed to this.

The Alberta researcher was present for the focus group conversations and prompted and observed the teachers' conversations about teaching. Hence, the researcher was an observer of the focus group and also part of the focus group.

Ontario Anglophones

Four Grade 7 teachers, each with at least six years of teaching experience, from four different schools, formed the Ontario focus group. We refer to them here by the pseudonyms Ellen, Jordan, Trista, and Steve. Three of their four schools include all grades from Junior Kindergarten up to Grade 8, while Jordan's school has only Grades 7 and 8. A range of in-class supports were available to the teachers, and the structure of schedules for teaching mathematics varied.

The researchers provided each teacher with a USB key containing each recorded lesson and then scheduled a follow-up meeting in which the teachers could inform the research assistant how to edit the 45–60-minute video of the lesson down to a 20-minute video clip. This clip would be shared with the other three teachers. The teachers took this step very seriously and would often have detailed notes about where to stop and start the edits. The researchers watched the entire lesson with each teacher. As the teachers identified the edit points, they would talk about why they were making that particular edit, for instance, to focus on a critical question that a student had asked, to follow a particular group throughout the lesson, or to demonstrate the variety of ways in which students approached a problem. These discussions were unexpectedly rich and seen by the teachers as an opportunity to reflect on their teaching, but unfortunately they were not recorded. More details of the data collection can be found in Chapter 3.

Quebec Anglophones

The anglophone focus group in Quebec consisted of three teachers, who will be referred to by the pseudonyms Tamara, Camilla, and Kayla. Camilla and Kayla, when the research began, had been teaching at the same school for at least seven years. A fourth teacher from the same school withdrew from the project for medical reasons during the first phase of the project, but gave us permission to use her videos even though she only participated in a couple of focus groups. The rate of diagnosed learning difficulties in the school is high. The teachers were former colleagues of one of the research assistants (Terry) and were recruited directly. This prior relationship helped to create a safe environment built on trust and respect. The three teachers taught at Secondary Cycle 1 in Quebec (Grades 7–8). Tamara and Camilla recorded their videos in their Grade 7 classes; Kayla recorded hers in a Grade 8 class of students with learning difficulties.

Terry facilitated the auto-ethnography discussion in which the teachers selected which of their own videos would be shared with teachers in the other regions. Another researcher was also present and mainly facilitated the ethno-ethnography discussion in which they watched videos of teachers from other regions of Canada.

The researchers offered to meet with the teachers to involve them in editing the videos, but the teachers were comfortable leaving the editing to the research assistant. None took up the invitation to review the edited videos before they were shown to the focus groups. More details of the data collection can be found in Chapter 4.

Quebec Francophones

The francophone focus group in Quebec consisted of four teachers, teaching in different schools within the same school board. They were selected by a mathematics school board consultant who approached them to participate in the study. His choice was made on the basis of the teachers' innovative practices and their interest in knowing more about teaching practices. The teachers will be referred to by the pseudonyms Carine, Irène, Johanne, and Roger. Johanne was in a program in which each of her students had access to a laptop computer. Roger taught in a low-income school that had an international student program. Irène also taught in a low-income school, but this school had special programs in music and in science. She had a group with many students who had learning difficulties in mathematics. Carine was teaching in a low-middle-income school.

As with the anglophone group, the researchers offered to meet with the teachers to involve them in editing the videos, but the teachers were comfortable leaving the editing to the research assistant. Only Irène took up the invitation, as she wished to review one of the edited videos before it was shown to the focus groups. She preferred that the problematic behavior of some students not be shown on the video. More details of the data collection can be found in Chapter 4.

New Brunswick Francophones

The francophone focus group in New Brunswick consisted of four teachers from three school boards. They were recruited by one of the research assistants (Dominic) who knew them from previous research projects. The teachers will be referred to by the pseudonyms Anne-Marie, Chloé, Danielle, and Guillaume. Anne-Marie and Chloé taught different grade levels in the same school. Anne-Marie used a flipped classroom in mathematics. Danielle taught at a technology-oriented school, and her students had access to laptops in her classroom. Danielle and Guillaume were at schools in remote locations.

In New Brunswick, the teachers recorded their lessons themselves with the help of a student from the class or a teaching assistant, although they were offered the help of a research assistant. During the editing process, Chloé suggested that the research assistant keep one part of her lesson in the edited version.

As in Quebec, most of the teachers did not feel the need to become involved in editing the video, nor to review the edited videos before they were shown to the focus groups. More details of the data collection can be found in Chapter 4.

Research Assistants in Quebec and New Brunswick

The data collection and analysis for these three focus groups was handled by the research group at McGill University. This included eleven research assistants: two of them were completing their Bachelor of Education degrees, three of them were doing their master's degrees, and six of them their doctoral degrees.

CONCLUSIONS

Our goal was to compare teachers' pedagogies across regions of Canada. Within our group, we debated the extent to which this was an achievable, or even an advisable, goal. It is clear that a focus group cannot represent a population, and it would be a mistake to claim that the pedagogies of a few teachers from a region are the same as those of all teachers from that region. This is a criticism that can be made of case studies in general. Tobin, Wu, and Davidson (1989) address this issue.

> Clearly, one preschool cannot be assumed to represent the preschools of a nation. … To address the problem of typicality and to give a sense of the great range of preschool practices and beliefs to be found in each country, we showed our videotapes to audiences associated with other preschools in the same country, in other cities. … In each city we asked preschool parents and staff, and students and faculty at university early child education programs, to tell us in what ways the preschools we videotaped were typical or atypical of others in their country. … These responses to our videotapes provide a third voice in our book, a narrative strain that contextualizes both our videotapes and the insiders' explanations of the tapes. (p. 8)

In our research, the selection of videos by teachers from one region to share with teachers from the other regions was carried out in similar ways. Except for the anglophone Quebec teachers, who were mostly from one school, the teachers in the focus groups taught in different schools and had rarely or never met before the focus group sessions. Hence the videos which they selected, while still not definitively typical (after all, there may be no such thing), were probably not extremely atypical. In any case, for the purposes of our research our main concern was with the teachers' reactions to the videos, not to the representativeness of the videos themselves.

So, while we do not claim the focus groups are representative, we are willing to consider them typical, in that their pedagogies do not fall outside the range one might encounter within their region. For the reasons previously outlined, we hypothesize that there is more similarity within regions than across regions. As we assume pedagogies influence teaching practice and teaching practice influences learning, we expect that the differences we see in these teachers' pedagogies could be reflected in differences in student achievement. Our results are only suggestive, however, and much further research would be necessary to establish this link.

In the chapters of this book, we explore the range of ways we used to gain some insight into teachers' pedagogies by examining the data from the focus group sessions, and some of the issues that arose during our analyses. In keeping with the enactivist methodology of the project, we did not seek a common perspective and approach to the data analysis; this is reflected in the diversity of approaches described in the remaining chapters. By adopting multiple and sometimes contradictory perspectives, we reveal more than can be seen from a single perspective.

Our observations of the teachers acted as inspiration for other explorations. As we went along, the project opened up new questions, which are described and discussed in subsequent chapters. These questions emerged through the interactions of each researcher with the data, and hence, they allowed for new research in directions we had not anticipated. These new explorations are as much an outcome of the project as anything we might now be able to say about regional pedagogies. In Chapter 11, I reflect back on the outcomes and the project as a whole.

REFERENCES

Anderson, J., Rogers, T., Klinger, D., Ungerleider, C., Glickman, V., & Anderson, B. (2006). Student and school correlates of mathematics achievement: Models of school performance based on pan-Canadian student assessment. *Canadian Journal of Education, 29*(3), 706–730.

Anderson-Levitt, K. M. (2002). *Teaching cultures: Knowledge for teaching first grade in France and the United States.* Cresskill, NJ: Hampton Press.

Beaton, A. E., & O'Dwyer, L. M. (2002). Separating school, classroom, and student variances and their relationship to socio-economic status. In D. F. Robitaille & A. E. Beaton (Eds.), *Secondary analysis of the TIMSS data* (pp. 211–231). Dordrecht, the Netherlands: Springer.

Brochu, P., Deussing, M-A., Houme, K., & Chuy, M. (2013). *Measuring up: Canadian results of the OECD PISA study.* Toronto, Canada: Council of Ministers of Education, Canada (CMEC). Retrieved from http://cmec.ca/Publications/Lists/Publications/Attachments/318/PISA2012_CanadianReport_EN_Web.pdf

Brown, L., & Reid, D. A. (2006). Embodied cognition: Somatic markers, purposes and emotional orientations. *Educational Studies in Mathematics, 63*(2), 179–192. doi: 10.1007/s10649-006-9027-3

Brown, L. C., & Reid, D. A. (2004). Emotional orientations and somatic markers: Implications for mathematics education. In M. J. Høines & A. B. Fuglestad (Eds.), *Proceedings of the Twenty Eighth Conference of the International Group for the Psychology of Mathematics Education* (Vol. 1, pp. 123–126). Bergen, Norway: PME.

Bruner, J. (1996). *The culture of education.* Cambridge, MA: Harvard University Press.

Bussière, E., Cartwright, F., Crocker, R., Ma, X., Oderkirk, J., & Zhang, Y. (2001). *Measuring up: The performance of Canada's youth in reading, mathematics and sci-*

ence: OECD PISA study-first results for Canadians aged 15. Ottawa, Canada: Statistics Canada.

Chazan, D., & Herbst, P. (2011). Challenges of particularity and generality in depicting and discussing teaching. *For the Learning of Mathematics, 33*(1), 9–13.

Chazan, D., Herbst, P., & Sela, H. (2011). Instructional alternatives via a virtual setting: Rich media supports for teacher development. In O. Zaslavsky & P. Sullivan (Eds.), *Constructing knowledge for teaching secondary mathematics: Tasks to enhance prospective and practicing teacher learning* (pp. 23–37). New York, NY: Springer.

Council of Ministers of Education, Canada. (1997). *School achievement indicators program (SAIP) 1997 Mathematics II assessment.* Toronto, Canada: Author. Retrieved from: http://www.cmec.ca/Publications/Lists/Publications/Attachments/138/saip1997.en.pdf

Herbst, P., & Chazan, D. (2006). Producing a viable story of geometry instruction: What kind of representation calls forth teachers' practical rationality? In S. Alatorre, J. L. Cortina, M. Sáiz, & A. Méndez (Eds.), *Proceedings of the Twenty Eighth Annual Meeting of the North American Chapter of the International Group for the Psychology of Mathematics Education* (Vol. 2, pp. 213–220). Mérida, Mexico: Universidad Pedagógica Nacional.

Herbst, P., & Chazan, D. (2011). On creating and using representations of mathematics teaching in research and teacher development. *ZDM—The International Journal on Mathematics Education, 43*, 1–5.

Lieberman, A. (Ed.). (1988). *Building a professional culture in schools.* New York, NY: Teachers College Press.

Maturana, H. (1987). Everything is said by an observer. In W. I. Thompson (Ed.), *Gaia: A way of knowing: Political implications of the new biology* (pp. 65–82). Great Barrington, MA: Lindisfarne Press.

Maturana, H. R. (1988). Reality: The search for objectivity or the quest for a compelling argument. *The Irish Journal of Psychology, 9*(1), 25–82.

Maturana, H. R. (2002). Autopoiesis, structural coupling and cognition: A history of these and other notions in the biology of cognition. *Cybernetics & Human Knowing, 9*(3–4), 5–34.

Reid, D. (2004). What is Canadian about Canadian mathematics education? In E. Simmt & B. Davis (Eds.), *Proceedings of the 2004 Annual Meeting of the Canadian Mathematics Education Study Group / Groupe Canadien d'Étude en Didactique des Mathématiques* (pp. 157–163). Edmonton, Canada: CMESG/GCEDM.

Schmidt, W. H., McKnight, C. C., Houang, R. T., Wang, H. A., Wiley, D. E., Cogan, L. S., & Wolfe, R. G. (2001). *Why schools matter: A cross-national comparison of curriculum and learning.* San Francisco, CA: Jossey-Bass.

Simmt, E. (2018). Curriculum in Canada: A fractal interpretation using the case of Alberta. In D. R. Thompson, M. A. Huntley, & C. Suurtamm (Eds.), *International perspectives on mathematics curriculum* (pp. 103–130). Charlotte, NC: Information Age Publishing.

Tobin, J. J. (1989). Visual anthropology and multivocal ethnography: A dialogical approach to Japanese preschool class size. *Dialectical Anthropology, 13*, 173–187.

Tobin, J., Hsueh, Y., & Karasawa, M. (2009). *Preschool in three cultures revisited: China, Japan, and the United States.* Chicago, IL: University of Chicago Press.

Tobin, J. J., Wu, D. Y. H., & Davidson, D. H. (1989). *Preschool in three cultures: Japan, China, and the United States.* New Haven, CT: Yale University Press.

Van Manen, M. (2008). Pedagogical sensitivity and teachers' practical knowing-in-action. *Peking University Education Review, 1*, 1–23.

Wilkins, J. L. M., Zembylas, M., & Travers, K. J. (2002). Investigating correlates of mathematics and science literacy in the final year of secondary school. In D. F. Robitaille & A. E. Beaton (Eds.), *Secondary analysis of the TIMSS data* (pp. 291–316). Dordrecht, the Netherlands: Springer.

PART 1

INTRODUCTION TO PART I

In a study that involves multiple phases of data collection and a full team of researchers, there are a number of different data sets that can be investigated by the various researchers. Part I contains chapters that share two qualities: first, they focus on the data that were gathered from the teacher participants who met with other teacher participants in their province to view the video recordings of their own teaching; and second, they were written by the *insider* researchers, those who conducted the research with the teacher participants. More specifically, the chapters in Part I are observer interpretations of features of pedagogy based on what the teachers discussed about their own teaching and the teaching of the other teachers as they viewed video recordings of themselves and others in their classrooms.

The analysis and writing of each chapter were done by the researcher(s) from the region where the teachers taught. We consider them insider researchers as they recruited the teacher participants, video recorded the teachers teaching, worked with the teachers to produce the edited videos the group of teachers viewed, facilitated the viewing and discussions of the video-recorded lessons, and followed through with all the other components of the research such as transcription, analysis, and the writing of the chapters in this part of the book. As is reinforced throughout this book, everything is said by an observer (Maturana, 1987). Hence, it is not surprising that this insider perspective of observing teachers observing their own teaching resulted in four chapters that each provides a distinct discussion of pedagogy.

Researching Pedagogy and Practice with Canadian Mathematics Teachers, pages 23–25.

In Chapter 2, we hear of the research that Elaine Simmt facilitated with the Alberta teacher participants. Elaine is an Albertan, and except for a very short period of time when she lived in British Columbia, she has lived in Alberta her whole life. Elaine is a former school teacher of mathematics, chemistry, and physics and has been working with teachers and curriculum leaders for some 25 years. Although her own education and teaching experience have been located in Alberta, she was the director of two international teacher development projects. Since 2010, she has worked in Tanzania on projects that ask how universities can work with communities (professional and civic) to build capacity for mathematics teaching in rural communities. In the last 4 years, she has been working with teacher educators in Oman to develop teacher training modules for in-service teachers. That work has become a site for researching international collaborative work. Working internationally has provided Elaine with an opportunity to reflect on curriculum and pedagogy as grounded in the local context. It is not surprising that her work in this project took a turn to a more theoretical question about the nature of the notion of pedagogy itself. In Chapter 2, Simmt explores multiple interpretations of pedagogy that are expressed in the educational literature in relation to the discussion of the Alberta teachers. In particular, she investigates their conversations about engagement.

Chapter 3 is a report of Christine Suurtamm and research assistant Brenna Quigley's work with the four Ontario teachers. Christine Suurtamm began her career as a secondary school mathematics teacher; as she moved into academia, her research has focused on mathematics teachers' classroom practice as they interact with students' ways of thinking about mathematics. In particular, much of her work has focused on teachers' formative assessment practices as they question, listen to, and respond to students' thinking. Much of her research involves working alongside multiple stakeholders, such as teachers, policy makers, and those responsible for teacher professional development. These interactions help develop a sensitivity to the constraints and affordances of this variety of contexts that support student and teacher learning. Chapter 3 emphasizes the multiple levels of noticing that occurred in this project—teachers noticing in the context of teaching, teachers noticing as they observe their own and others' teaching, and researchers noticing what teachers notice. In the chapter, they discuss the lesson components and environments that the teachers saw as conducive to developing mathematical thinking and understanding.

In Chapter 4, the members of the Quebec research team, Annie Savard, Dominic Manuel, Sophie Pinard, Scosha Merovitz, and Vandana Chandrasekhar, discuss their work with three teacher focus groups: two teacher groups, one anglophone and the other francophone, from Quebec; and one francophone group from New Brunswick. Annie Savard, the lead researcher in the Quebec team, is a former elementary school teacher, having taught for 15 years at that level. She has also worked with the Quebec Ministry of Education focusing on supporting elementary school teachers in regard to teaching mathematics in elementary school.

She has a master's degree in science education and a doctorate in mathematics education. She obtained her degrees in French, and she has been a faculty member at McGill University since 2007, where she has developed a bilingual section of two mathematics education courses for pre-service elementary school teachers. She is also involved in projects, funded by the World Bank, on numeracy in early grades in different African countries. Her research program focuses on how teaching and learning mathematics in school contributes to developing citizenship among students. This research program includes financial education and the use of technology, and it has a focus on determining how to support teachers in teaching mathematics to support students to develop their reasoning skills. In Chapter 4, the researchers draw upon the bilingual nature of the team by leaning on two different cultural frameworks in their data analysis with elements from French *didactique.* See, e.g., Brousseau's (1998) and Lampert et al.'s (2013) ambitious pedagogies.

In Chapter 5, three members of this same research team return to the discussions of their three teacher groups. In this chapter, they focus on teacher perceptions of their own and other's teaching practices in the context of teaching fractions.

In reading these chapters, the reader may wish to consider:

- In what ways do teachers in different provinces appear to have some similar views of pedagogy? Different views?
- When considering that "everything is said by an observer" (Maturana, 1987), how might the researchers, as observers of the teachers, be seeing things in different ways?
- Some of the chapters suggest that the teacher discussions provide insights into what the teachers value. Are there ways that these chapters also provide insights into what the researchers value?

REFERENCES

Brousseau, G. (1998). *Théorie des situations didactiques* [Theory of didactical situations in mathematics, 1970–1990]. Grenoble, France: La Pensée Sauvage.

Lampert, M., Franke, M. L., Kazemi, E., Ghousseini, H., Turrou, A. C., Beasley, H., . . . Crowe, K. (2013). Keeping it complex: Using rehearsals to support novice teacher learning of ambitious teaching. *Journal of Teacher Education*, *64*(3), 226–243.

Maturana, H. (1987). Everything is said by an observer. In W. I. Thompson (Ed.), *Gaia, A way of knowing: Political implications of the new biology* (pp. 65–82). Great Barrington, MA: Lindisfarne Press.

CHAPTER 2

PEDAGOGY AS ENGAGEMENT

Elaine Simmt

The purpose of this chapter is to explore notions of pedagogy as inferred from comments that teachers made when viewing video recordings of their own and other teachers' lessons. In this chapter, one case of three teachers from Western Canada is explored. This group of teachers placed a great deal of emphasis on engagement. That emphasis emerged from their discussions about the strategies they use in their practice, their conceptions of their role as teacher, and their intentions for their students. Their conversations about engagement are explored in relation to conceptions of pedagogy offered in the literature, and in relation to the broader discourse in education in their region.

What might we hear if we ask teachers to talk about their pedagogy? In Canada, the response might include, among others, a description of their instructional approach, a philosophy of their teaching, their orientation towards curriculum and instruction, or possibly their ways of working with children. But, what might we infer about their pedagogy if, instead of asking them about their pedagogy, we listened for indications of it by asking teachers to comment on lessons taught by themselves and others? A unique opportunity to explore various conceptualizations of pedagogy was provided in the project as teachers observed themselves and others teaching. In this chapter, I use different conceptions of pedagogy to interpret discussions among three teachers from the same school system in Alberta, Canada commenting and reflecting upon video recordings of lessons from their own classroom and the classrooms of others. This chapter, then, is both about the

Researching Pedagogy and Practice with Canadian Mathematics Teachers, pages 27–41.

pedagogy of the teachers and about how particular conceptions used to interpret emergent themes lead to different discussions about the teachers' pedagogy. In particular, the notion of engagement, which emerged in the discussions among these teachers, is taken up and interrogated using multiple nuances of the conception of pedagogy.

CONCEPTUALIZING PEDAGOGY

The use of the word *pedagogy* is far less common in English texts than the words *teaching* or *instruction* (NGRAM, August 17, 2018). Using Google books Ngram viewer to analyze frequency suggests that the use of the word *pedagogy* in English texts has had two remarkable growth periods, the first between 1900 and 1920, and the second in the last three decades. Indeed, since its usage last peaked in the 1910s, current usage has more than doubled in the 2000s. A search in JSTOR of *Educational Studies in Mathematics* and *Journal for Research in Mathematics Education* for the words *pedagogy*, *pedagogical*, and *pedagogue* revealed only 30 papers with one of these words in the title. No paper titles included the word *pedagogue*, and the most common usage was in the phrase *pedagogical content.*

A contemporary dictionary definition of pedagogy—"the method and practice of teaching, especially as an academic subject or theoretical concept" (*Google Dictionary*, n.d.)—does not reflect the depth of meaning it has in the field of education, nor does it reflect historical understandings derived that might be inferred from its etymology. A pedagogue in Greek references was responsible for the upbringing of the child, a pedagogue was more of a custodian than a tutor (Smith, 2006). The etymology of pedagogy reveals its connection to 16th century middle French that suggests pedagogy is teaching (Harper, 2020). In contemporary usage, pedagogy is more closely associated with the 16th century French definition than the ancient Greek usage (Smith, 2006). However, the nuances of its conceptualization vary within different educational traditions and among researchers even within the same tradition. In some of the literature, the term *pedagogy* is used interchangeably with the terms *teaching* or *instructional strategies* (Anthony & Walshaw, 2009). In other literature, pedagogy is about the enacted teacher-learner relationship (van Manen, 2008). In yet other schools of thought, pedagogy is more about teaching situated within context and discourse that includes educational theories, beliefs, policies, and issues (Alexander, 2008). Three perspectives of pedagogy are offered here to illustrate how attention can be focused quite differently in an exploration of a teacher's pedagogy. The first is pedagogy as focused on teaching content; the second is pedagogy as a relationship between teacher and student; and the third is pedagogy as teaching performance and the discourses that inform it within a context.

Anthony and Walshaw (2009), in *Effective Pedagogy in Mathematics*, equate pedagogy with teaching. The authors use the phrase *effective pedagogy* in their title; however, the content of the book reveals discussion about effective mathematics *teaching*. The authors write: "This booklet *focuses on effective mathemat-*

ics teaching [emphasis added]. Drawing on a wide range of research, it describes the kinds of *pedagogical approaches* [emphasis added] that engage learners and lead to desirable outcomes" (p. 6). The pedagogical approaches the authors propose include:

- Caring for learners and caring about their development in mathematics (p. 7);
- Providing students with opportunities to make sense of ideas independently and collaboratively (p. 9);
- Planning lessons to build on students' proficiencies, interests, and experiences (p. 11);
- Understanding the relationship between the tasks and examples used to promote student understanding (p. 13);
- Supporting students in making connections within the content, and between the content and other dimensions of their experience (p. 15);
- Using a variety of assessment practices to observe student thinking and to support student learning (p. 17);
- Facilitating meaningful and purposeful classroom dialogue (p. 19);
- Facilitating language development within the content domain and communicating with learners in ways they can understand (p. 21);
- Selecting tools and representations to support student meaning making (p. 23); and
- Using sound content and pedagogical content knowledge of the discipline (p. 25).

Anthony and Walshaw ground these pedagogical approaches in beliefs about the child's right to education, capacity to develop mathematical identities, respect and sensitivity to the child's heritage, thinking processes and realities of the classroom, optimization of a range of educational outcomes, and enhancement of a range of social outcomes that contribute to the holistic development of students for participation in society (p. 6).

Simon's (1995) constructivist perspective on pedagogy is aligned with Anthony and Walshaw's interpretation of pedagogy insofar as his primary concern is with the teacher's understanding of content and learners. In his effort to reconstruct mathematics pedagogy from a constructivist perspective, Simon focuses on teaching as contextual, planned, and lived. He illustrates pedagogy in terms of a teaching cycle that incorporates teachers' knowledge of mathematics, mathematics activities and representations, hypothetical learning trajectories, and teachers' hypothesis of student knowledge, and their theories of mathematics learning. This conception of pedagogy is nuanced by Wager (2010) in her study of equitable mathematics pedagogy. She places emphasis neither on particular content, nor on a teaching cycle, but rather she focuses on understanding how teacher actions can create an environment for participation in mathematics.

Shifting our gaze from pedagogy in terms of the teaching of academic knowledge and know-how for teaching particular content, van Manen (2008) asks us to think about pedagogy in terms of the relationship between teacher and child. He returns to the origins of the meaning of pedagogy:

> The concept of 'pedagogy'– the study or practice of guiding or rearing children — has long carried the meaning of discretion, prudence, judgment, caution, forethought (van Manen, 1997) .... Moreover, the concept of teacher as pedagogue assumes that he or she is motivated by a caring interest in the growth and welfare [of] children. In other words, teaching is not only governed by principles of effectiveness, but also by special normative, ethical, or affective considerations. (p. 6)

It is from this position that van Manen understands pedagogy to include a teacher's practical knowing and sensitivity to the learner, including the relational, personal, moral, and emotional aspects of working with children and youth (p. 5). Such a focus results in observing practical knowing actions and sensitivity of the teacher, rather than a focus on strategies or practices.

Alexander (2008), in contrast to having an inward focus as van Manen (2008) does with his emphasis on the teacher-student relationship, focuses bidirectionally, explicitly including the context in which teaching takes place. He describes pedagogy as both "an act of teaching and the theories, beliefs, policies and controversies that inform and shape it" (p. 3). He draws attention to a continental view of pedagogy, which includes "both the act and the idea of teaching" (p. 5). In this view, pedagogy is distinguished from didactics, that is, the science of teaching a particular subject (p. 5). (See Chapter 4 for further discussion about didactics.) By invoking Alexander's conception, pedagogy is explored in relation to the teacher and learner in terms of the multiple discourses in which they are embedded.

The overarching study, from which this chapter emerged, used the following working definition of pedagogy: the largely implicit, underlying "set of beliefs, tacit knowledge, customs and habits that guide practice" (Reid et al., 2015, p. 368). In this chapter, rather than focus on that definition, I extend the project's conceptualization of pedagogy to include a discussion of multiple interpretations grounded in teachers' reflections on lessons, as interpreted through the three views of pedagogy noted previously. Observing teachers observing mathematics lessons taught by themselves and others led me to explore three Alberta teachers' discussions about teaching. From those discussions, the notion of engagement emerged. This led me to ask how their desire for learners to engage may be observed as a feature of each teacher's pedagogy. I ask:

- In what ways does a pedagogy of engagement speak to teaching practices?
- In what ways does a pedagogy of engagement contribute to the pedagogical relationship?
- In what ways is the teachers' pedagogy of engagement a part of the broader context of education?

OBSERVING PEDAGOGY

Regardless of one's conceptualization of pedagogy, all researchers face the same problem of determining ways to study pedagogy. This methodological problem was a challenge for the pan-Canadian research group, who intended to study how pedagogies in middle school mathematics in regions of Canada compare. Of particular significance was the project researchers' epistemological stance that "everything said is said by an observer" (Maturana, 1987, p. 65). Hence, the observations a researcher makes speak to the researcher's experience of the observed rather than to an objective perspective of the observed phenomenon. In this work, rather than observing teachers teach, or interviewing them about their teaching, the project was designed to provide opportunities for teachers to observe themselves and others teaching and engaging in conversations with colleagues about the teaching they witnessed. The researcher was present for the conversations, and in the Alberta case, prompted and observed the three teachers' conversations about teaching (referred to as focus groups), and systematically reviewed transcripts of the conversations. Hence, the researcher was an observer of the focus group, part of the focus group, and subsequently an observer of the observing (study of transcripts). In this way, the discussion also speaks about the researcher's views of pedagogy.

The study procedure involved video recording three mathematics lessons taught by each of the three teachers in the study. The teacher participants then gathered to view their own videos and to select an exemplary lesson and a typical lesson, from among the nine lessons, to share with teachers in two other Canadian provinces. Once videos of lessons were selected, they were shared with teachers from other regions in Canada. The Alberta teachers then met, viewed the videos from teachers in other regions, and talked about teaching in other parts of Canada. Videos of the Alberta teachers' discussions of both selecting videos to send to other teachers and viewing the videos of other teachers were the data for this study (see Chapter 1 for details about the overarching study). Although the videos of their own teaching were to be the primary prompts for the first discussion, I (as the researcher) opened the discussion with the teachers by asking a general question, "What do you think are three elements or features of *your* pedagogy that make a difference to learners?" Throughout the discussions, I inserted statements that repeated or rephrased a comment (which I reflectively call *prompts*). For example:

Teacher: We do a lot of working together. And I do a lot of problem solving.

Researcher: And so another one is a lot of problem solving.

A second type of comment pointed to what I was interested in having the teachers explore.

Researcher: Okay, so those are some of the things that we can kind of think about, because it is interesting that two of you immediately said, sort of, working together, math talk, ... that kind of thing. So then I have, um—yeah, I guess my question was, would you,—do you think that other teachers are focusing on math talk in their classroom.

A third form of exchange had me inserting my views:

Teacher: Because I have a lot of kids that are fully engaged in participating when I'm doing a lesson, and then it comes to work time, and they just shut right down on me. And without that practice, even though they appear to get it, they just, without the practice they're lose—they're just not getting it, right, they're gonna lose it all.

Researcher: That's why I like this little chart (pointing to a poster on the classroom wall that lists Kilpatrick's[1] elements of mathematical proficiency), because it says yes—we need procedural fluency, right? It's something that they think we lost a little bit when sort of constructivist things came in and everybody was, "Okay, it's all about one thing instead of about all of these things."

Transcripts from the discussions were coded for utterances and comments that the researcher understood to be related to pedagogy. A set of preliminary categories was created to reduce the data. Attention was paid to the comments and questions made by the teachers that reflected surprise, puzzlement, curiosity, concern, or some other form of emphasis. It seemed such comments might reveal a teacher's implicit beliefs, tacit knowledge, customs, and habits, as those practices, actions, and interactions within a lesson with which a teacher is most familiar are unlikely to generate such responses. Expressions reflecting familiarity were also noted. I then took the categories and examined them against the initial conceptualization of pedagogy as the beliefs, tacit knowledge, customs, and habits that guide practice to produce inferences about the teachers' pedagogy; these are reported as themes. Codes were created through a recursive process of reading the transcripts for potential indicators of pedagogy as understood in our working definitions and developing themes from those codes. Finally, the themes were analyzed in broader terms as I sought to reflect on pedagogy from multiple perspectives. These themes can be compared with those observed by other members of the research team who studied the transcripts (see Chapters 6, 7, 8).

[1] Kilpatrick, J., Swafford, J., & Findell, B. (Eds.). (2001). *Adding it up: Helping children learn mathematics*. Washington, DC: National Academy Press.

THE TEACHERS AND THEIR PEDAGOGY

Three teachers (Brenda, Maryanne, and Andrea) reflected on and discussed teaching prompted by both videos of their own and others' teaching and the comments and questions I inserted into the research sessions (see above for examples). The teachers all taught for a large urban school district; however, their schools were situated in different communities and each teacher brought her own history of professional experience to the classroom. In their discussions, comments about the context in which they teach emerged. Very brief information about the teachers and the schools in which they taught is presented in Figure 2.1.

A set of emergent categories was extracted from the coding of transcripts of teachers talking about their own lessons and observing the lessons of others. From those categories, further interpretive work resulted in a set of themes that can be discussed in terms of the teachers' pedagogy (see Table 2.1): engagement, teacher-learner interaction, felt constraints, and teaching approaches. It seemed, from the teachers' comments and reflections, that the desire to engage students in lessons, in mathematics, and in school shaped many of the decisions and actions that the teachers took. This desire for engagement could stem from teacher beliefs that engaged students are learning and are making something out of the time they spend in class and in school. The teachers were also observed to focus on what I have labeled as *felt constraints*. Felt constraints are factors that permeate pedagogy but are believed to be out of a teacher's immediate control.

Brenda had experience as a district mathematics consultant and teaches mathematics in an inner city junior high school (Grades 7–9) that is situated in one of the city's oldest neighborhoods— a neighborhood that sits at the intersection of two of the lowest average income communities in the city. For years, the school population has been declining. Transience and absenteeism are among the challenges teachers and administrators experience at the school. Brenda commented about one of her challenges: "*You have to keep them on board—so you can get through everything.*"
Maryanne was working on her Master of Education degree at the time of the study. She too teaches in a school that is understood as inner city by the local school authority. It is located in a mature mixed housing neighborhood with single-family dwellings, some townhouse developments, and a few apartments. It is one of the most affordable neighborhoods in terms of home ownership and rental costs. Many of the children's families are new and second generation in Canada. The school offers both specialized programming in English language learning and advanced placement in the school. Absenteeism can be a problem. Maryanne commented about her learners, "*It's actually getting the kids to do their work.*"
Andrea had been a department head. She teaches in a junior high school in a neighborhood with mixed housing and a median house price above the city's average. The population is diverse, but traditionally the neighborhood was home to people employed in skilled and professional work. The school has a program for high-achieving students in addition to offering core programming. Andrea framed her challenge as, "*I like my kids to take risks, and be willing to take risks, and struggle.*"

FIGURE 2.1. Participants' Background and Teaching Context (pseudonyms used)

TABLE 2.1. Categories and Themes from which Pedagogy Is Inferred

Category	Locus of Action	Emergent Theme
Working together Problem solving Practice Making sense Active	Students	Engagement
Math talk	Students & teacher	Engagement
Checking for understanding Attending to student responses	Teacher	Teacher-learner interaction
Textbooks Classroom arrangement Topics of instruction	Teacher	Felt constraints
Chunking activities Problem posing Direct teaching	Teacher	Teaching approaches

In their discussions, the teachers also discussed checking for student understanding and responding to student questions. I have separated these from engagement because the categories here were less about a desire for engagement and more about the teacher focusing on checking in on the student's understanding of mathematics itself.

Finally, the themes were read against the three understandings of pedagogy found in the literature. The themes can be discussed in relation to the teachers' beliefs about mathematics teaching and learning, their relationships with students, and the situational and cultural context in which they work. The theme of engagement will be used to discuss multiple interpretations of pedagogy.

A Desire for Engagement

One of the most prominent themes found in the data from this group of teachers was that they all expressed a desire for learners to be engaged. Each teacher made explicit comments in this regard. Brenda said that she tries "to keep them engaged in whatever we're doing." Maryanne commented on a fun activity that connects with the kid's life: "it's just to engage them." Andrea admitted that most of her energy goes into how she is going to engage the kids. She said, "How am I going to make them interested enough to learn this well?" Each teacher expressed a desire to engage learners, but their motivations and approaches to achieve this goal were not the same.

Brenda expressed an explicit need to keep the learners "on board" so she could get through "everything." She is the teacher who faces transience and absenteeism issues in her classes. Brenda claims:

> If I am not engaging those kids, they don't bother coming to math. So I have to be doing my song and dance every day for them. Or else, they'll go down the road to the mall and get their poutine— I have to.

When viewing her own video, which had been edited to a 20-minute segment, Brenda was concerned that the edited video did not show how she interacted with the learners. Her concern for this absence suggests that she values a pattern of instruction that the missing segments contained, as suggested in the following:

> Well, I think that notion of chunking the activities where we're doing a little bit of problem solving, and then getting the kids—giving the kids a chance to do it, and then having that walk-around time, which was chopped from our videos, right?

She was bothered that the "bit where I [she] was walking around to see if everyone was *with me* [her] is gone" [emphasis added].

This notion of the students being with her suggests a desire for engagement. How might this be interpreted pedagogically? How does the teacher know the students are with her? Does her noticing the missing piece of the lesson point to that dimension of pedagogy that van Manen (2007) describes as pathic?

> The competence of professional practitioners is itself largely tied into pathic knowledge. Professional knowledge is pathic to the extent that the act of practice depends on the sense and sensuality of the body, personal presence, relational perceptiveness, tact for knowing what to say and do in contingent situations, thoughtful routines and practices, and other aspects of knowledge that are in part pre-reflective, pre-theoretic, pre-linguistic. (van Manen, 2007, p. 20)

Is the teacher seeking in the videos a representation of her presence and relational perceptiveness to the learners? What might she mean by *with me*? An initial interpretation could be that the teacher interacts with the students so they stay engaged—pay attention—stay on task. A second interpretation of *with me* is that the learners understand the mathematics that she is teaching. Brenda remarked at one point, "Always answer a question with another question." It seems the comment about being *with me* could mean both things to her—the learners stay on task, and they understand what she is doing mathematically. Further, to keep the learners engaged Brenda chunks her activities. Considering this strategy of chunking activities within the context of some of her other comments suggests that she designs lessons that enable her to get students to come to class, to ensure they are on task when in class, and to closely monitor the students for their understanding while on task.

Of the three teachers, Maryanne seemed to focus more on "actually getting the kids to do their work." Her efforts to keep students engaged differed from Brenda's. Maryanne commented, "Just ask the questions until I kind of get the response I'm looking for." She felt that this was a means of guiding the learners. This strategy seems similar to Brenda's, "answer a question with another question." However, the strategy of asking until the teacher gets the response she is looking for suggests a funnelling approach to questioning (Bauersfeld, 1988); this was not reflected in Brenda's comments. Whereas some forms of questioning can lead students to finding relationships and making connections, other forms of questioning, such as funnelling, simply lead students to a correct answer. Maryanne explained her motivation for using this kind of questioning: to get students "to think through it ... by pointing them in the right direction instead of telling them." A funnelling questioning strategy could be seen as a form of teacher-learner interaction that keeps learners engaged in the moment; it may be used by the teacher believing it is helpful to the student, but it may result in learner dependency on the teacher (Yackel & Cobb, 1996). The desire to keep students engaged could be about the immediate interactions within the classroom, or it could refer to engagement with mathematics beyond the classroom.

Andrea, too, was concerned that the videos, edited from videos of her complete lessons, did not show the students working on the mathematical activities she prepared for them. However, her comment seemed to express different intentions than Brenda and Maryanne had expressed. Andrea said:

> I think that ... it's about what the students are bringing. So, that idea that they're contributing. Which I think ... we don't see a lot in the videos, because that's [what happens] more [in] their [student] work time [which isn't reflected in the video].

Andrea reflected on beginning her classes with problems for the students to solve. She noted that sometimes those problems took up the whole class even though she may have planned for them only as a starter (as it did in one of the recorded lessons). Furthermore, the problems Andrea posed were intended to engage the learners in mathematical thinking and practices—not simply to get them to do the work, or to motivate their engagement with context from their everyday lives. She expressed a desire for her learners to "take risks, and be willing to take risks, and struggle." From her point of view, doing mathematics involves "a lot of struggling." She understood engagement in mathematics teaching and learning quite differently from the other two teachers, who broke things down and simplified them in order to keep the learners engaged. For Andrea, engagement in mathematics was not doing the work, but working on something challenging. "You know," she said, "you can train them, and they do get comfortable with it." But that was not good enough for her students. Her teaching included helping students understand that working on something challenging is worthwhile. "It's okay to say in front of everyone something that might not be correct, and we'll go from there... just try something." van Manen (2008) notes that:

> At the end of the day, what matters to many teachers is that they could provide their students with positive experiences, that there was a good atmosphere in their classes, that students felt safe and successful in their learning activities, that personal difficulties could be worked out, that life that day was happy or good for them and their students. (p. 5)

Pedagogy as Desire for Engaged Learners

Recall the perspective from which this research is conducted, it is the observer (Maturana, 1988) who listens and interprets the teachers' comments, and subsequently explains those comments as teachers desiring engaged learners. The teachers spoke about the lessons that they teach and that they observed in the videos. I observe and explain their comments as a desire, an emotional orientation for learner engagement (Drodge & Reid, 2000; Maturana, 1988). This emotional desire, both tacitly and explicitly, emerged in the teachers' conversations as they reflected and explained their classroom actions and interactions. In the overarching study in which these teachers participated, we began with a working definition of pedagogy: a set of beliefs, attitudes, and orientations to teaching. It was using this definition that the desire for engagement emerged in the initial analysis and interpretation of the data. However, other definitions of pedagogy are available for re-interpreting the theme of engagement. Three additional definitions were proposed in the literature review: pedagogy as a teacher's strategies and approaches to teaching; pedagogy as a relationship between the teacher and student; and pedagogy as both act and discourse of teaching. In the context of this chapter, I ask how a pedagogy of engagement is reflected in these different perspectives.

Engaging Learners through Teaching Strategies and Approaches

Returning to Anthony and Walshaw's (2009) commentary on pedagogy, we can see how pedagogy as engagement can be manifested in various teaching practices, from planning lessons to build on student proficiencies, to using student interests and experiences, to having students make sense of ideas independently and collaboratively. Over the course of their discussions, the teachers in this study commented on the structure of their lessons: using particular questioning techniques; connecting mathematics to life outside of school; offering challenging problems to begin a class; and designing games to keep students motivated. The teachers discussed how they arranged their classroom seating plans to encourage math talk and to facilitate learning, and they noted that their style of teaching keeps them "on their feet" as they walk around the classroom deliberately monitoring student work and posing questions to prompt further student actions and meaning making. These examples provide some insight into understanding pedagogy as the teaching of content.

Engagement through Relationship

Next, I ask in what ways is a pedagogy of engagement a reflection of van Manen's (2008) conceptualization of pedagogy. From his perspective, teaching is not about effectiveness, but about the normative, ethical, and affective considerations of the teacher for the student that are manifested in a relationship where the care of the student is paramount. The relationship between teacher and student, which emerges from a pedagogy of engagement, is grounded in the care that the teacher has for the student. A sense of teaching as efficiency is reflected in the teachers' conversations about the strategies they use to keep students engaged. However, we also heard the teachers making connections with the students' day-to-day lives through their choice of examples, understanding of their lives outside of school, and thinking about some of the overarching life strategies they want their students to take away from their lessons. This can be interpreted as a teaching strategy, but it can also be thought of in terms of caring for the learners (Hackenberg, 2005; Long, 2011), and developing relationships with them.

Van Manen (1994) suggests that pedagogy is the relationship between teacher and pupil. This relational aspect of pedagogy was not part of the original definition of pedagogy used in the research. However, in exploring the theme of engagement, the relationship between teacher and learner becomes salient. In the case of the mathematics class, the teacher's care is expressed in her selection of meaningful tasks, her encouragement of mathematics talk, her checks for understanding, and her attention to student response. Understanding and care create the opportunities for creating and nourishing the pedagogical relationship.

A Discourse of Engagement

Interpreting engagement by altering the conception of pedagogy to take an outward look provides yet more insights. In what ways is the desire for engagement as pedagogy embedded in the socio-cultural and political forces within the context in which these teachers teach? The teachers themselves recognized that the students they teach, the content they are teaching, and the community in which they teach all shape how they work with the learners in their classrooms. Given the differences in the teachers' three school communities, I was reminded of the work of Anyon (1980), who found a hidden curriculum of work (preparation) in four distinct socio-economic class contexts: two working class schools, one middle class school, one affluent professional school, and an elite school. Although the teacher discussions in this study did not reflect the extremes that Anyon found in her study—preparing the students from the working class neighborhoods for factory jobs, the middle class for skilled work, and the affluent professional for more academic work, and the elite for problem solving—there was a marked difference between what engagement meant to Andrea compared to Brenda and Maryanne. Broadly speaking, Andrea taught in a more affluent neighborhood and focused on students taking risks and challenging themselves, whereas Brenda and Maryanne were focused on keeping students on task.

Why did the notion of engagement emerge in their discussions? Why not *participation* in mathematics (Wager, 2010), for example? Is participation in mathematics the same as engagement? Alexander's (2008) discussion of pedagogy suggests that we might look to the broader social and political context in which the pedagogy of these teachers is embedded, including curriculum, policy, teacher education—the professional, political, and public educational discourse.

In particular, the notion of engagement is found in educational materials for mathematics teachers. Indeed, Anthony and Walshaw's (2009) monograph, cited in this paper, uses precisely that term. Phrases such as engaging learners and engagement permeate papers in professional journals, such as those published by the National Council of Teachers of Mathematics. Engagement is also noted as a theme in workshops and other professional development activities in which these teachers participate. When a particular emphasis, like engagement, is placed on teaching and learning in professional development contexts, teachers encounter that emphasis in the social domain both directly, by participating in those activities, and indirectly, as their peers bring the emphasis back to their schools.

The policy discourse in Alberta also includes an emphasis on engagement, from the overarching goals of education, to statements in curriculum policy documents. For example, the goal of education in Alberta is to:

> Inspire all students to achieve success and fulfillment, and reach their full potential by developing the competencies of *Engaged Thinkers* [emphasis added] and Ethical Citizens with an Entrepreneurial Spirit, who contribute to a strong and prosperous economy and society. (Government of Alberta, Ministerial Order (#001/2013))

The provincial curriculum (program of study) for mathematics in Alberta notes that students who meet the goals of mathematics education will "engage and persevere in mathematics" (Alberta Education, 2007/2016, p. 3). Teachers are advised: "Creating an environment where students openly look for, and engage in, finding a variety of strategies for solving problems empowers students to explore alternatives and develops confident, cognitive mathematical risk takers" (p. 6).

Professional journals, professional development conferences and workshops, educational policy statements, and curriculum documents all perpetuate a discourse of engagement in which teachers are immersed. Unpacking pedagogy in this broader context of professional and political discourse adds yet another dimension to understanding teacher pedagogy.

CONCLUSION

As we try to understand pedagogy, an appropriate starting place is with the observable in teachers' lessons, and in the teachers' reflections on those lessons. The three teacher participants commented about teaching, learning, and their schools as they observed videos of their own teaching and the teaching of others. One of the themes that emerged from the study was engaging students. In order to understand pedagogy, I unpacked this theme in light of three different conceptions of pedagogy: as teaching, as relationship, and as shaped by educational, social, and political discourses. The teachers in this focus group talked about the strategies they use in order to encourage student engagement in their lessons and in mathematics. In their comments, the teacher-student relationship could be observed in the deliberate choices they made to understand their learners as individuals and as youth. Finally, by observing their reflections in the broader context of professional development, the curriculum documents, and educational policy, we also get insight into why a pedagogy of engagement emerged from the discussions of these three Alberta junior high school mathematics teachers.

Although this analysis of teachers' comments on videos of teaching was done for only one group of teachers, other analyses in this book suggest that the data from any of the provinces in this study, using multiple conceptions of pedagogy, could result in yet a deeper understanding of pedagogy itself. Reflexively conceptualizing pedagogy in multiple ways provides a more complex interpretation of teachers' (largely implicit) underlying sets of beliefs, tacit knowledge, customs, and habits that guide practice.

REFERENCES

Alberta Education, Government of Alberta. (2007/2016). *Mathematics: Kindergarten to grade 9 program of studies*. Retrieved from: https://education.alberta.ca/media/3115252/2016_k_to_9_math_pos.pdf

Alexander, R. (2008). Pedagogy, curriculum and culture. In K. Hall, P. Murphy, & J. Soler (Eds.), *Pedagogy and practice: Culture and identities* (pp. 3–27). London, UK: Sage.

Anthony, G., & Walshaw, M. (2009). *Effective pedagogy in mathematics*. Geneva, Switzerland: International Bureau of Education. Retrieved from: http://www.iaoed.org/downloads/EdPractices_19.pdf

Anyon, J. (1980). Social class and the hidden curriculum of work. *The Journal of Education, 162*(1), 67–92.

Bauersfeld, H. (1988). Interaction, construction, and knowledge: Alternative perspectives for mathematics education. In D. A. Grouws & T. J. Cooney (Eds.), *Perspectives on research on effective mathematics teaching* (pp. 27–46). Reston, VA: National Council of Teachers of Mathematics.

Drodge, E. N., & Reid, D. A. (2000). Embodied cognition and the mathematical emotional orientation. *Mathematical Thinking and Learning, 2*(4), 249–267.

Google Dictionary. (n.d.). Pedagogy. Retrieved from: https://www.google.com/search?q=pedagogy&oq=pedagogy&aqs=chrome..69i57j46j0l6.2007j0j8&sourceid=chrome&ie=UTF-8

Government of Alberta. (2013). Ministerial Order (#001/2013) Student learning. Retrieved from: https://education.alberta.ca/media/1626588/ministerial-order-on-student-learning.pdf

Hackenberg, A. (2005). A model of mathematical learning and caring relations. *For the Learning of Mathematics, 25*(1), 45–51.

Harper, D. (2020). *Online Etymology Dictionary*. Pedagogy. Retrieved from: https://www.etymonline.com/word/pedagogy

Long, J. S. (2011). Labelling angles: Care, indifference and mathematical symbols. *For the Learning of Mathematics, 31*(3), 2–7.

Maturana, H. (1987). Everything is said by an observer. In W. I. Thompson (Ed.), *Gaia, A way of knowing: Political implications of the new biology* (pp. 65–82). Great Barrington, MA: Lindisfarne Press.

Maturana, H. R. (1988). Reality: The search for objectivity or the quest for a compelling argument. *The Irish Journal* of *Psychology, 9*(1), 25–82.

NGram Viewer (2018, August 17). https://books.google.com/ngrams/graph?year_start=1800&year_end=2008&corpus=15&smoothing=7&case_insensitive=on&content=pedagogy&direct_url=t4%3B%2Cpedagogy%3B%2Cc0%3B%2Cs0%3B%3Bpedagogy%3B%2Cc0%3B%3BPedagogy%3B%2Cc0%3B%3BPEDAGOGY%3B%2Cc0

Reid, D. A., Simmt, E., Savard, A., Suurtamm, C., Manuel, D., Lin, T. W. J.,…Knipping, C. (2015). Observing observers: Using video to prompt and record reflections on teachers' pedagogies in four regions of Canada. *Research in Comparative and International Education, 10*(3), 367–382.

Simon, M. A. (1995). Reconstructing mathematics pedagogy from a constructivist perspective. *Journal for Research in Mathematics Education, 26*(2), 114–145.

Smith, M. J. (2006). The role of the pedagogue in Galatians. *SOR Faculty Publications and Presentations*, 115. Retrieved from: https://digitalcommons.liberty.edu/sor_fac_pubs/115

van Manen, M. (1994). Pedagogy, virtue, and narrative identity in teaching. *Curriculum Inquiry, 24*(2), 135–170.

van Manen, M. (2007). Phenomenology of practice. *Phenomenology and Practice, 1*, 11–30.

van Manen, M. (2008). Pedagogical sensitivity and teachers practical knowing-in-action. *Peking University Education Review, 1*, 1–23.

Wager, A. A. (2010). Teacher positioning and equitable mathematics pedagogy. In M. Q. Foote (Ed.), *Mathematics teaching and learning in K–12: Equity and professional development* (pp. 77–92). New York, NY: Palgrave.

Yackel, E., & Cobb, P. (1996). Sociomathematical norms, argumentation, and autonomy in mathematics. *Journal for Research in Mathematics Education, 27*(4), 458–477.

CHAPTER 3

NESTED NOTICING

Valuing Voice

Christine Suurtamm and Brenna Quigley

This chapter focuses on the observations of four Grade 7 teachers in Ontario who shared about teaching mathematics through their reflections on their own and one another's teaching. The main research questions were: What do teachers think are significant components in teaching? What do they value in teaching? Three layers of noticing are presented, including (1) teachers noticing student actions within their own classrooms, (2) teachers noticing teacher actions as they sit with colleagues and observe episodes of teaching, and (3) researchers noticing what teachers notice and pay attention to. Data consisted of video-recorded discussions of four teacher participants as they watched and discussed videos of their own teaching and of each other's teaching. Through a thematic analysis of the data gathered, we created a concept map that presents these teacher practices, what the teachers seemed to value, and the teachers' goals in teaching mathematics. The teachers' overarching goal that emerged through their conversations was promoting mathematical thinking and understanding.

This chapter focuses on our work of analyzing the collected data to understand what teachers value about mathematics teaching and learning and how these values emerge in pedagogical practice. We share the analysis of the data we gathered from working with four Grade 7 Ontario anglophone mathematics teachers across two years. In year 1, we worked with the teachers to video record three lessons from each of the four teachers. At the end of year 1 and into year 2, we met with

Researching Pedagogy and Practice with Canadian Mathematics Teachers, pages 43–59.

the teachers on several occasions as a group to view and discuss their lessons. Thus, in this study, we worked with four teachers, and together we viewed samples of their teaching. The teachers commented on their own and one another's teaching in ways that tended to highlight what they seemed to think are important components of teaching mathematics. Because the teachers were drawing on their own video-recorded practice, they anchored their views in practice. What is different about this study is that, rather than teachers talking about teaching in an interview or a focus group, in these sessions, the researchers used videos of the teachers' teaching to prompt the discussion and for the teachers to comment on aspects of the viewed lessons. A great deal of the focus in the research literature on mathematics teaching tends to be on shifting teaching practice. That was not our intention in this study. Rather, we wanted to view teacher practices from the teachers' perspectives. Through an inquiry stance (Cochrane-Smith & Lytle, 2009), we sought to understand teachers' views of important mathematics teaching practices, and to understand some of the values that underlie teachers' decision making and their choice of practices to use within the classroom context.

NESTED NOTICING

"Every act of teaching depends on noticing" (Mason, 2002, p. 7). Teachers are constantly paying attention to student actions and understandings. Quite often this is what guides a teacher's next steps. Teachers watch and listen to what students do and they act accordingly. Jacobs, Lamb, and Philipp (2010) emphasize that noticing is subjective, and teachers do not all notice the same things in the same ways. They suggest that noticing can be defined as "(a) attending to children's strategies, (b) interpreting children's understandings, and (c) deciding how to respond on the basis of children's understandings" (p. 169). This is what we consider to be the first layer of noticing.

A second layer of noticing occurs in this study, which involves teachers observing their teaching and the teaching of others and noticing or paying attention to teacher actions. Paying attention to their own moves in practice is often difficult for teachers to do as they are busy with the day-to-day business of planning and teaching, and often their classroom practices become routine and may be left unquestioned (Mason, 2002). In this study, through working in a group and watching videos of teaching, teachers have the opportunity to focus on particular aspects of their own and others' teaching. The use of video provides time to reflect on the complexity of a teaching moment or environment and to unpack the variety of things that can be noticed and discussed (Walkoe, 2015). The teachers would often stop a video they were watching together, and a discussion would ensue regarding a particular teacher move. In many cases, they picked out moments where the actions of the teacher in the video mirrored some of their own actions or intents. The discussions that followed focused on particular aspects of teaching, and the conversations that evolved included what they were doing as well as why they were doing it. We suggest that these conversations get at what teachers notice, and, hence, what they value.

A third layer of noticing pertains to what we, as researchers, notice about what the teachers notice. Mason (2002) suggests that much of qualitative research is about noticing and determining what moments within research settings are significant. This third layer of noticing is the bulk of the work that qualitative researchers do; it is important work as, in this context, it brings to life the work of teaching and what teachers value.

To summarize, we see three layers of noticing in this inquiry: (1) teachers noticing student actions within their own classrooms, (2) teachers noticing teacher actions as they sit with colleagues and observe episodes of teaching, and (3) researchers noticing what teachers notice and pay attention to. In the next section, we outline the process we used to investigate these nested layers of noticing.

RESEARCH PROCESS

Our research design helped to reveal teachers' practices as well as their views and perspectives (Reid et al., 2015). When teachers were invited to participate in this study, we stated that we would be focusing on being able to describe what mathematics teaching looks like in Canada, and in this case, in Ontario in particular. We also explained to teachers that we wanted to understand why teachers made particular pedagogical moves. Our focus was not on what teachers did in their classrooms as much as the discussion they had when they collectively viewed their own and one another's lessons. Thus, as with all of the research reported in this volume, although we video recorded four teachers' lessons, the video recordings of the lessons were not our data. Our data comprised the video-recorded discussions that the four teachers had while they sat and watched the video recordings of their own and one another's lessons.

Four Grade 7 teachers volunteered to participate in this study: Ellen, Jordan, Trista, and Steve. These teachers were all from the same school district, and the invitation to participate was shared with all Grade 7 teachers in their schools. The four teachers who participated all stated that they were interested in research in education and that they were curious about what it would be like to be a research participant. They also stated that they hoped the research would provide them with a professional learning opportunity. As outlined below, each teacher taught in a different school and had different types of prior teaching experiences.

Ellen had six years of teaching experience at the time of the study and teaches in a school that includes Kindergarten to Grade 8. In this school, Grade 7 and 8 students are on a rotary system where the students have different teachers for different subjects. Ellen taught both math and science. For this study, the video recordings were of her mathematics class that included both Grade 7 and 8 students. She differentiated her lessons in a variety of ways to meet the needs of students in both grades.

Jordan had 12 years of teaching experience at the start of the study and was teaching in a Grade 7/8 urban school. Jordan taught both mathematics and social studies. The video recordings were of his Grade 7 class, which was a large class with over 30 students. There were often two to three support staff or educational assistants present in the room to assist individual students.

Trista had 8 years of teaching experience, including two years supporting professional learning through the Curriculum Services department of her school board. At the time of this study, she taught at a Kindergarten to Grade 8 urban/suburban school serving a population that includes many recent immigrants to Canada, with students representing over 40 countries and languages. The Grade 7 class that was video recorded had 20 students. Trista taught these students all of their subjects except for music and French. Mathematics was taught in the morning as Trista found that students were most attentive at this time.

Steve had been teaching for 11 years, 10 of which were spent teaching mathematics and science. At the time of this study, he taught in a Kindergarten to Grade 8 suburban school that was the largest of the four schools involved in the Ontario study. He taught nearly all subjects to one class of students, but he also taught mathematics to one other class, a class of 20 Grade 7 students, which was the class that was video recorded.

In the first phase of our study (see Phase 1 in Figure 3.1), we video recorded three lessons taught by each of these four Grade 7 teachers. In this phase, the teachers were included in both the research and the decision making in a variety of ways. First, we asked the teachers to invite us to three lessons that they wanted to have recorded. Depending on the school structure, the recorded lessons ranged from 45–70 minutes. After each lesson was recorded, each teacher worked with the researcher and the research assistant to edit each video into a 20-minute video clip that could be viewed by the others. The instructions they were given were to retain the parts of the original video that would help highlight the key characteristics of the lesson and of their teaching. As the teachers worked with us in moving from full classroom videos to edited videos, they made choices about what to retain in each of their three 20-minute videos. Our role was one of supporting the teachers in the technical aspects of viewing the full video and recording the portions they wanted to include in the edited video. In our work, the teachers were fully in charge of what they wanted to share, and they made choices based on a variety of factors, such as wanting to share episodes that:

- Focused on student-to-student interaction;
- Followed a particular group of students through a lesson;
- Kept a framing piece, such as the establishment of a certain mathematical model;
- Provided examples of the teacher interacting with small groups;
- Shared a strategy (teaching or a student solution) that they deemed unique; and/or
- Shared a teacher struggle.

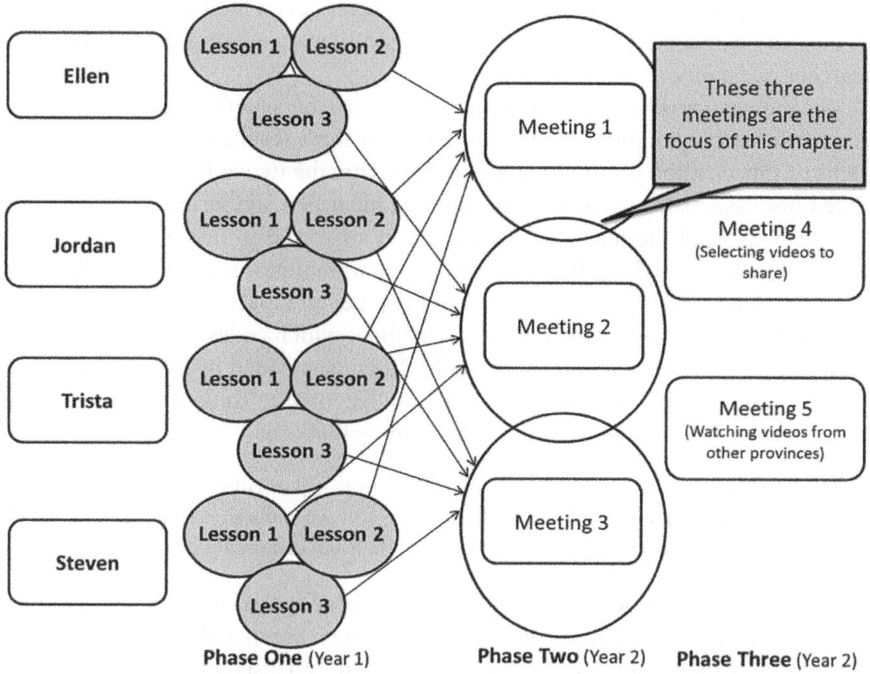

FIGURE 3.1. Phases of Data Collection

It should be noted that this process of having the teachers make the decisions about what the edited video looked like was not necessarily consistent with what was done in other provinces. However, we saw this as an important component to ensure that the video represented what the teacher valued. At the end of this process, we then had twelve 20-minute videos (i.e., three videos from each of the four Grade 7 teachers' classrooms).

The second phase consisted of three focus group (FG) meetings with the four teachers. During each of these meetings, the teachers observed and discussed four classroom video recordings, one from each of the teachers. These group meetings were video recorded. The analysis of the data from this second phase helped us see what practices the teachers notice in their teaching, and consequently, through the discussions that ensued about their practices, what the teachers value about mathematics teaching. We were interested in teachers' descriptions and reflections on their own practices first and foremost; however, we were also interested in what we could learn from the teachers as they shared their thoughts on teaching as a product of viewing other teachers' practices. Ultimately, we sought to answer the questions: What do teachers think are significant components in teaching? What do they value in teaching? Within each of the group meetings, we focused on making space for the teachers to think and to reflect on their teaching. Because of this, the conversations occurred naturally, rather than being prompted by researcher questions.

The thematic analysis of the data gathered as teachers met to view the videos of their own teaching and the teaching of the other three teachers in the group is the focus of this chapter. For our analysis of this data, the two authors independently coded discussions in which the teachers had identified similarities between their own teaching and the teaching practices they observed in one another's video-recorded lessons. We both began by making margin notes in transcribed datasets, and we independently created codes. We then met and discussed these codes and their meanings, and we came to a decision about which codes we would use. For instance, the following comment made by Ellen was coded by one researcher as "student talk" and by the other researcher as "listening to student thinking." Ultimately, we decided on the code: "students share thinking."

> So, I wanted them to be doing a lot of talking in that particular lesson and I think they did a lot of discussion of what they knew, so I was able to hear all of that and then know where to go from there in terms of their understanding. (FG1)

We then worked independently again using these agreed-upon codes, and then we discussed and re-coded the discussions. We then compared the way we coded the data and worked through the discrepancies through discussion (Jordan & Henderson, 1995). We noticed that there were different types of codes we were using. Although some codes were related to particular teaching practices or lesson components, when the teachers commented on the similarities between the teaching practices, the discussion often also focused on the rationale for using particular practices, which thus, provided insights on what it was about the practice that the teacher(s) valued. For instance, in the excerpt above, and others that were coded as "students share thinking," we saw a connection being made with what the teachers valued. We grouped the coded data under particular headings, such as classroom environment or teacher practices, but we also sought to see the relationships between the types of coded data, or rather, the headings.

RESULTS

In the sections that follow, we present the results of our analysis of the discussions from the first three focus group sessions in Phase 1. We were challenged to think of ways to organize and report on our data analysis, because categories overlapped and could fall into several organizational structures that included teacher practices and teacher values. In refining our analysis, we developed a concept map (see Figure 3.2) that represented the organization of our data analysis into the following categories: teacher practices, what teachers seemed to value, and the overall goals in teaching mathematics (as defined by the participants). The teacher practices were those lesson components that the teachers noticed in their own

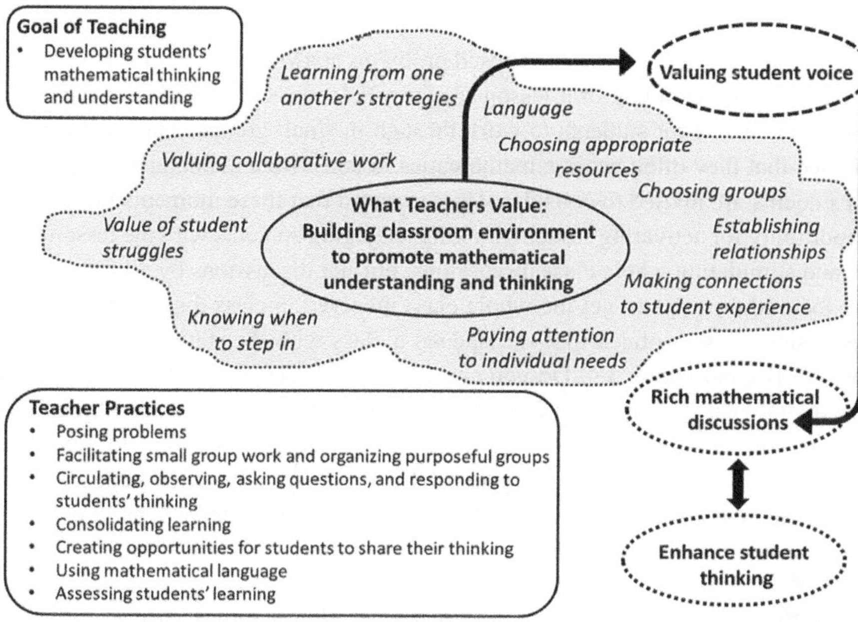

FIGURE 3.2. A Representation of Our Data Analysis That Shows Teacher Practice, What These Teachers Value in Building Mathematical Understanding and Thinking, and the Teachers' Overall Goals of Mathematics Teaching.

and one another's teaching. We noticed that what the teachers seemed to value was often expressed during the conversations where the teachers had extensive discussions about particular teaching practices and during the conversations when they talked about the classroom environment they were trying to create. These were things we noticed. The overarching teaching goal that emerged through the teachers' conversations was promoting mathematical thinking and understanding. We also found that, related to this goal, the teachers were noting that they value student voice and rich mathematical discussions.

TEACHER PRACTICES

The teachers recognized many similarities in one another's lessons, and they focused on some of those or on teacher practices when discussing the video-recorded lessons (see bottom left of Figure 3.2). Some of the similarities included posing problems, presenting opportunities for small group work, circulating and observing as students work, as well as asking questions and responding to students' thinking. The teachers also noticed the different ways they created space for whole class discussion, consolidating ideas, and encouraging students to share their thinking.

Posing Problems

The teachers noticed that they posed problems at a variety of points in a lesson, such as at the beginning of a lesson for the whole class to consider, or part way through a lesson for students to work through in small groups. The teachers explained that they often begin a mathematics lesson with a mathematical problem all students are invited to consider. They reported that these moments provide an opportunity for activating student thinking, engaging students with the lesson topic, and stimulating whole class discussions. Further discussions by all four teachers focused on ways to get the whole class involved in class discussions. In one discussion, Trista noticed that she opened a class with the statement: "Jeff says that 2/3 is greater than 3/4. Do you agree with Jeff? Yes or no? And be prepared to explain what you think." (FG 1). Jordan responded that he will sometimes say, "Okay, how many people agree with so and so?" (FG 1) as a way to stimulate student thinking. The teachers suggested that these questions encourage students to consider the mathematical concepts, and they prompt students to practice communicating their thinking with clarity and precision.

Facilitating Small Group Work and Organizing Purposeful Groups

The teachers noticed that they all create opportunities for small group work in their mathematics classes. They expressed that they want to create an environment that allows students to work through their thinking, to try out different methods, and to be able to share those methods with other students. For instance, Steve explained that for one of his lessons students were working in small groups and each group was given a different graph to work with. After each group had worked with their graph for a while, they shared their work with the class, and everyone had the opportunity to see the different ways that students were interpreting the graphs.

As part of the discussion on collaborative learning, the teachers discussed how the formation or selection of purposeful groups occurs in their classes. The teachers said that they sometimes use random selection, while at other times, they select particular groupings to ensure there are mixed ability groups. Trista noted that she "change[s] [her] groups all the time" and sometimes students work in a "mixed ability group situation" (FG 1). Ellen also noted that, in her class, "depending on the task, they might be in ability groups, but more often than not it's mixed ability" (FG 1). She added:

> And it's, sometimes, it's their choice, other times it's mine, especially with a lot of partner or pair work, I'll just let them pick who they work with—pair or group of three, go ahead stand up, find somebody to talk with, but with the seating groups, it's generally my choice. (FG 1)

Circulating, Observing, Asking Questions, and Responding to Students' Thinking

When talking about the small group work, the teachers often commented on the conversations they had with students. The teachers emphasized that while students are working in small groups, the teachers are able to circulate, observe, ask questions, and respond to students' thinking in different ways. Ellen noted an instance where she thought one student group was struggling because they had introduced additional complexity into the problem. She said that this "wasn't a bad thing in the end because they had lots of time to get through it. And that shows the importance of making sure that we're walking around and we're asking those questions, we're observing what they're doing, challenging their thinking" (FG 2). Ellen explained that she was asking "guiding questions" that might help the groups think about the methods they were using or how they were progressing.

As well as looking at student thinking, the teachers also used this opportunity to observe the group dynamics. During the third focus group session, Steve shared a comment that highlights how attuned the teachers are to the learning and contributions of individual members of the group. Reflecting on the video excerpt he shared, Steve noted that one student "sort of took that group over" (FG 3). Steve continued:

> You probably noticed that that was the first group that was finished. They were finished quickly, and he sort of was dominating the group. ...There were lots of students in the group that also understood it, it's just that he dominated it. (FG 3)

The teachers highlighted how they made these forms of informal assessment opportunities during small group work, as well as in other contexts, to ensure that each individual student understood the concepts being studied. Ellen, for instance, explained that when she observes mistakes or evidence of possible misconceptions in solutions, she takes note of the misconceptions on the clipboard she carries with her. She either addresses these by recapping a lesson with the full class, if there are several students who seem to be misunderstanding something, or she works individually or in small groups with particular students.

Consolidating Learning

The teachers noticed that they all tended to have whole class discussions to consolidate learning but they occurred in different ways. For Trista, consolidation often took place near the end of a class when students would all bring their chairs together to form a large group at the front of the classroom, and through discussion, they would "anchor" what they learned during that lesson. Trista stated that having students explain their thinking to the full group was not something that she was very successful in facilitating, though she stated: "That's still a piece that I'm working on" (FG 1). Jordan also spoke about consolidating learning; however, he pointed to an example from one of the lessons he shared in which he used consolidation at the beginning of a lesson. He referred to an excerpt in which his class took time to recap what students had discovered in previous classes through their investigation. He referred to this practice as a "mini-consolidate and a redirect at the same time" (FG 2). While Jordan felt that this approach "comes across as very dialectic" (FG 2), he felt that it was an essential "check-in point" or a "move ahead point—like a crossroads, if you will" (FG 2). He saw this as a moment for "reaffirming" what students have been learning (FG 2).

Creating Opportunities for Students to Share Their Thinking

The teachers spoke about how they provide students with many opportunities to share their thinking throughout each lesson. In the opening conversation of the first focus group session, Jordan noted that he believes that the "whole talking piece, and just listening to each other is really huge" (FG 1). When commenting on the lessons they were sharing and observing, the teachers highlighted moments where students were communicating and attempting to explain their mathematical thinking through interactions that were student-to-student(s), student-to-class, and sometimes student-to-teacher. Steve said, "I want them to be confident in what they know and be able to show me that they can clearly explain what they know, and be able to explain it to their peers" (FG 1). Commenting on one of the lessons she shared, Ellen also said:

> I wanted them to be doing a lot of talking in that particular lesson, and I think they did a lot of discussion of what they knew, so I was able to hear all of that and then know where to go from there in terms of their understanding. (FG 1)

Creating opportunities for students to share their thinking was a teacher practice that the teachers noted had value for both the students and the teacher, in particular, because the teacher used these opportunities to assess students' thinking.

Using Mathematical Language

The teachers noted that in their lessons they tried to model the use of appropriate mathematical language in their classes, and they also expressed an interest in paying attention to students' mathematical language as well. On one occasion, the teachers exchanged the following comments illustrating some of the ways in which they think carefully about how language is used in the mathematics classroom and what they might be doing themselves to build an environment where language use matters:

Trista: I found just the language that they were using was really interesting—that they were able to use precise language. It's not "Stuff," it's not "Things." They're using real words that mean something.

Ellen: Yes, and part of that is for us as teachers to expect that in the classroom, and if they're not we prompt them.

Jordan: Yeah, that's very important. I find myself saying [to them], "Well these two numbers when you put them together ...." I will say, "Well what do you mean? Do you mean products? Do you mean factors? What are we talking about here?" And so, you're getting them to specify and practice that language. (FG 1)

Later, Ellen stated to her colleagues: "And so you're getting them to specify and practice that language, and it breaks down those barriers . . ." (FG 1).

Commenting on one of the teacher's video-recorded lessons the group had just watched, Trista noticed that Steve, in recapping a lesson, modeled the use of mathematical language:

Trista: At the beginning when you were kind of recapping, I noticed the ... and it flowed very easily from you, but it was very ... I don't know if academic is the right word, but all the math language was flowing out of you very naturally, which I thought was really good for them to hear. Like it ... talking about adjacent and, you know ... [others saying "mhm"]. It just seemed really natural, which I thought was a good thing, because it just becomes part of the way that you speak.

Ellen: They pick it up.

Trista: And I want them to be able to speak in math that way, so I thought that was a great model to have.

Jordan: Yeah. As much as possible, I try to use that language. Generally, when I teach, I know I'm talking above a lot of them, but that's deliberate, because it means that they have to stretch, and then for the kids that you have to go back and rephrase for, then you go back and rephrase for, you'll do a visual for it. But I think that there's, you know, we often don't give children enough credit for their intelligence. They are very, very bright, much brighter than we realize, and if they realize that they have to ... "Adjacent, adjacent what's adjacent?" Then they're continually having to stretch and then think about, "Okay that means that ...."

Ellen: Yeah. And the student that asks you the question, about "Is it the same on the other side?" obviously didn't have that understanding of the word adjacent and he wanted to confirm it, so he took the risk and asked a question. So, there were probably 10 or 12 other kids in the room that had the same question, and that's awesome. Yeah.

Jordan: Yeah. So, we use a lot of the math language, like we'll talk, you know.

Trista: I think it's great how it just comes out as a part of the conversation. (FG 2)

Furthermore, establishing a "continuum" (FG 2) with how terminology is being used across the grades was mentioned throughout the teachers' discussions, and the teachers seemed to be conscious of how some concepts are discussed using different terms in elementary mathematics compared to secondary mathematics.

Assessing Students' Learning

The teachers noted that they seemed to assess students' learning in an ongoing manner. Throughout the focus group discussions, when commenting on students' developing mathematical thinking and understandings, the teachers seemed to emphasize, in particular, how important it is for each of them to get to know their students' individual needs as learners. The teachers shared a few strategies they used to help them assess students' mathematical thinking and understanding, and to help them develop a better understanding of each student's needs. Ellen, for instance, explained how she uses *exit cards*, on which students submit a short response to a prompt or question as they leave the class, as one strategy for verifying the extent to which students are developing an understanding of something presented during a class. Ellen noted that the information she collects through this short activity can provide her with information that can help guide her ongoing lesson planning for subsequent lessons. Ellen explains:

> I do use a lot of exit cards as well, in terms of individual student assessment, and that really helps because in a group task sometimes you think everybody gets it, and then you do the independent task and you realize, "Oh, there's actually half the class that isn't there yet." So, it definitely guides your instruction the next day. (FG 2)

Trista explained that she gives students an opportunity to self-assess their learning so they can state what they have learned and explain how they know what they know. Reflecting on her experience using this practice, Trista stated: "And you know that was the most interesting piece in the feedback I got? … [I]n that 'How do you know I'm right' section of it, the majority of answers were, 'I asked the teacher'" (FG 3). She continued to explain and Ellen added to the discussion:

Trista: Isn't that a great piece of feedback for me, that they need to learn their own strategies for doing it a different way, working backwards, whatever. You know, I mean, "Ask a teacher" and "Ask a friend" were the two different ones because they're always working together and I …. But they're relying on … I realized how much they're relying on my ….

Ellen: Outside feedback.
Trista: Yeah. And I don't want that, right?
Ellen: Yeah. (FG 3)

By giving students an opportunity to reflect on their learning, Trista came to realize that students were relying more on her, as the teacher, to confirm their thinking, rather than having students be able to verify their own solutions independently without external affirmation and be able to justify their responses confidently. The teachers commented numerous times on the importance of having students develop confidence in their knowledge and abilities. The teachers seemed to view their role as one that is supportive. Telling students whether or not they had the "right" answer was not the teachers' priority. Instead, the teachers seemed to value giving students opportunities to self-assess and to determine whether or not they could explain and justify their responses to the problems they were solving. Steve, for instance, stated:

> Right. So, it's like, "No I'm not going to tell you. However, if you get stuck, or you want to ask me a question that's going to maybe give you ideas or help guide you, I'll certainly answer that, but I'm not telling you what to do or how to do it. But I will help you." (FG 3)

WHAT THE TEACHERS SEEMED TO VALUE

When the teachers noticed aspects of their teaching practices in the video-recorded lessons and discussed their teaching practices, we observed that through discussion of these practices, the teachers seemed to be highlighting what they value in teaching mathematics. These values emerged in our coding and analysis of the data and are displayed in the shape in the center of Figure 3.2. In particular, the teachers emphasized that (a) establishing relationships, (b) paying attention and responding to students' individual needs, (c) identifying and working through students' struggles, (d) working collaboratively, and (e) making connections to students' experiences were all reflected in the practices they were highlighting. These were the ways in which the teachers seemed to be both supporting and engaging students in the mathematics classroom.

According to the teachers, an important part of constructing a classroom environment that is conducive to developing students' mathematical thinking and understanding is the relationships that exist in the classroom. Jordan, for example, noted that "there's a trusting kind of relationship" (FG 1) that forms in the classroom that becomes the foundation upon which students are able to share, respond, and also be challenged at times to participate and to think or rethink the responses and solutions they might be offering. Jordan explained how he uses time during whole group discussions, in particular, to ensure a lot of students are able to participate in discussions and that many solutions to the problems being posed can be shared. Jordan noted that managing these discussions is "a very complex task" (FG 1).

Paying attention and responding to students' individual needs also seemed to be highly valued by the teachers. This was reflected specifically in how they approached using ongoing and diverse forms of assessment in the mathematics classroom to learn more about how the students' mathematical thinking and skill development were progressing. Jordan also noted that observing interpersonal and intrapersonal factors was helpful when trying to learn how to best support a student's learning. When he is dealing with students who have more anxiety, for instance, and who might be seeking confirmation on whether or not their response is correct, Jordan suggested that he might say, "Yeah, you know, I think you're on track with that. That's pretty good. I like the way ..." and so on, to give them some reassurance (FG 3).

The teachers all agreed that making space for students to struggle and to be able to work through their thinking was something they valued in mathematics teaching. Steve shared the following thoughts:

> Trista: I like the struggle and I do that too—letting them struggle, letting them get frustrated, and when the student made that sound, that "arghhhh," I love that. Those are my favorite sounds.
>
> Jordan: The "I get it sounds." But, yeah, I feel that's a very important thing to let them do.
>
> Steve: Sometimes it's the "arrr" … then it's the "ahhh."
>
> Trista: Yeah, because we're so quick to help. And that's been a part of my own struggle as a teacher because you're so wanting to just get in there and help them and save them but the stepping back and giving them time and realizing that often they will get it and get more out of it when you let them do that—let them struggle. (FG 1)

The teachers all noted that knowing when to step in and intervene to help a student along was also important. They agreed that it depends on the needs of each student. The teachers' comments about how they support students' learning seemed to be an extension of their assessments of students' diverse needs.

In addition to the teachers supporting each student's learning, all teachers also emphasized that having students work together and discuss mathematics with one another was important. The teachers suggested that collaborative work could help students practice communicating their thinking and it could help students practice using language to clarify and share their understandings and to do so with precision. The teachers also suggested that collaborative work made it possible for students to have their thinking be challenged and/or be expanded by seeing other ways through which problems could potentially be solved and how they might be able to justify different solutions. Ellen and Steve both explained that they see value in having students get up and walk around the classroom to seek out support from their peers to help support their learning. Trista also commented on a similarity she observed across all four classrooms when she noted:

> I think I'm the only teacher in my junior and intermediate division [in my school] where students sit in groups … but what I see with all four of us is that our students are talking and working together, so we obviously all value that. (FG 2)

For Jordan, this collaborative dimension to their lessons and classroom environment aligns with what he expressed as the "notion [of] social construction of understanding or [the] social construction of knowledge," where students are working together to develop their thinking and construct their mathematical understandings (FG 2).

The teachers also discussed the significance of making connections to students' experiences and prior knowledge as a way of helping students develop their mathematical thinking and understandings. The teachers referred to both students' lived experiences, generally speaking, and also students' experiences learning mathematics when considering their learning within the present course and also their prior courses. Ellen, for example, spoke about one instance where students provided many sports examples to illustrate when they might encounter negative integers in daily life. Jordan suggested that students be reminded or that they remind themselves of what they already know when they return to concepts that they are expected to have worked with previously, which serves to credit students with some prior learning rather than presenting something as entirely new.

While analyzing the teachers' discussions about their teaching practices, we also felt that one of the values that seemed to resonate with all four of the teachers was valuing students' voice in the mathematics classroom. Although this was not a specific phrase that the teachers themselves introduced during their conversations, unlike the previous forms of engagement that are reflective of what the teachers seemed to value in their mathematics classrooms, positioning the student prominently as not only a participant but as a primary contributor to discussions and problem-solving processes in the mathematics classroom seemed to be important for all four teachers.

CONCLUDING COMMENTS

The practices the teachers noticed, such as posing problems, facilitating small group work and organizing purposeful groups, circulating, observing, asking questions, responding to students' thinking, consolidating learning, creating opportunities for students to share their thinking, using mathematical language, and assessing students' learning, all seemed to be contributing to learning environments that were conducive to developing mathematical thinking and understanding. Each of the four teachers also spoke about developing students' mathematical thinking and understanding when they spoke about their teaching practices.

Throughout this chapter, we presented instances of nested noticing, where we see noticing as an intentional act of paying attention to ways of thinking and doing (Anthony, Hunter, & Hunter, 2015). Nested noticing included the ways teachers noticed students' thinking, how the teachers noticed their own teaching practices, and how we, as researchers, noticed what the teachers seemed to value in regards to mathematics teaching. Similar to how the teachers paid attention to students' thinking as something we saw they valued in the classroom, we, as researchers, paid attention to and valued teacher voice. When preparing for the focus group sessions, we wondered and planned how we might prompt teachers in talking about their teaching. Our task was to set up the environment for their voices to be heard. During the focus groups though, few prompts were required other than to describe the process to them. The teachers would start talking as the video played and we would stop the video to allow the discussion to flow. The teachers saw the focus group sessions as a form of their own professional development; they enjoyed talking about teaching and seemed engaged. Our task as researchers was then to listen to the teachers, to notice the ideas that they wanted to talk about, to provide a risk-free environment for discussion, and to respond to their thinking. Even though the teachers had not met prior to our first focus group session, we were struck by the similarities the teachers were seeing in one another's teaching.

The ways in which this research was designed made it possible to create a space in which reflections on teaching and productive discussions were possible. Having teachers view segments of their teaching and discuss their teaching practices created opportunities for the teachers to consider many different aspects of the mathematics classroom environment and how they teach mathematics. Through the focus group sessions, the teachers were not only able to watch and reflect on their own classroom practices, but they were also able to consider their own teaching in light of the practices and values being shared by the other teachers. At the end of the second focus group session, the teachers commented on how much they appreciated having the time to reflect on and discuss their teaching practices with fellow teachers:

Researcher: I think we should let you go. It's quarter to nine. I don't mind staying longer. I just start to feel really guilty because I know you've got to drive home.

Trista: We love talking about . . . [Others are all agreeing and laughing. People are talking at the same time.]

Jordan: When we get together we're like minds. It's very addicting. It's just, I find …

Researcher: That's why we kept saying we couldn't wait to get you all together.

Jordan: I just, butt my head up against the wall so often. [Others are agreeing.] (FG 2)

At the end of the first focus group session, Trista commented:

> You know what, having the opportunity to talk with other teachers who are motivated to talk about teaching is something you don't get, I don't get, during the day very much, so I appreciate that time to get to talk about [teaching].

Ellen agreed, noting that she had also realized through her experiences working with a particular team at her school that "it's invigorating when you get to talk to people that are excited about [instructional practice]."

Through this research, we allocated both time and space to make it possible for teachers to reflect upon and share their thinking about teaching mathematics. Having the opportunity to observe, reflect upon, and discuss their teaching practices with other mathematics teachers seemed to be a productive experience for the participants in this study. We sought to create supportive spaces, much like the teachers created supportive spaces for students to develop mathematical thinking and understanding in their classrooms. As researchers, we listened to what the teachers shared about teaching mathematics, a reflection of our stance as researchers and what we value in research spaces.

REFERENCES

Anthony, G., Hunter, J., & Hunter, R. (2015). Supporting prospective teachers to notice students' mathematical thinking through rehearsal activities. *Mathematics Teacher Education and Development*, *17*(2), 7–24.

Cochrane-Smith, M., & Lytle, S. L. (2009). *Inquiry as stance: Practitioner research for the next generation*. New York, NY: Teachers College Press.

Jacobs, V. R., Lamb, L. L. C., & Philipp, R. A. (2010). Professional noticing of children's mathematical thinking. *Journal for Research in Mathematics Education*, *41*(2), 169–202.

Jordan, B., & Henderson, A. (1995). Interaction analysis: Foundations and practice. *The Journal of the Learning Sciences*, *4*(1), 39–103.

Mason, J. (2002). *Researching your own practice: The discipline of noticing*. New York, NY: Routledge.

Reid, D. A., Simmt, E., Savard, A., Suurtamm, C., Manuel, D., Lin, T. W. J., … & Knipping, C. (2015). Observing observers: Using video to prompt and record reflections on teachers' pedagogies in four regions of Canada. *Research in Comparative and International Education*, *10*(3), 367–382.

Walkoe, J. (2015). Exploring teacher noticing of student algebraic thinking in a video club. *Journal of Mathematics Teacher Education*, *18*(6), 523–550.

At the end of the first focus group session, Trista commented:

> You know what, having the opportunity to talk with other teachers who are motivated to talk about teaching is something you don't get. I don't get it during the day. Very much, so I appreciate that time to get to talk about [teaching].

Ellen agreed, noting that she had also realized through her experiences working with a particular team at her school that "it's invigorating when you get to talk to people that are excited about [teaching and practice]."

Through this research, we allocated both time and space to make it possible for teachers to reflect upon, and share their thinking about teaching moments. Having the opportunity to observe, reflect upon, and discuss their teaching practices with other mathematics teacher[illegible]

CHAPTER 4

OBSERVING TEACHERS WITH DIFFERENT LENSES

Where to Situate Them?

Annie Savard, Dominic Manuel, Sophie Pinard, Scosha Merovitz, and Vandana Chandrasekhar

This chapter presents different lenses to interpret observed teaching practices. Inspired by the French Didactics work of Brousseau (1998), we contrasted pedagogy and didactics to shed light on what teachers paid attention to. We contrasted teachers' attention on teaching practices that emphasized classroom management with their attention to teaching practices that support students to learn mathematical concepts and processes.

This chapter reports on three focus groups of teachers who participated in the study. Savard, as the researcher responsible for the two Quebec focus groups and the New Brunswick focus group, attended all focus group meetings. Manuel, a research assistant, usually accompanied her. Reflecting on these meetings, we were curious regarding the similarities and differences in the groups' observations of the videorecorded lessons, and we were interested to know more about the participating teachers' representations of teaching and learning mathematics. Our extensive data analysis and the clarification and refinement of our conceptual framework helped us interpret the observations made by these three groups of teachers. In this chapter, we share our findings.

Researching Pedagogy and Practice with Canadian Mathematics Teachers, pages 61–92.

Group	Region	Language	Participants and Grade Taught[1]
1 (QCA)	Quebec	English	Camilla—secondary 1 (12–13 years old)
			Kayla—secondary 2 (13–14 years old)
			Tamara—secondary 1 (12–13 years old)
2 (QCF)	Quebec	French	Carine[2] – secondary 1 (12–13 years old)
			Irène—secondary 1 (12–13 years old)
			Johanne—secondary 2 (13–14 years old)
			Roger—secondary 1 (12–13 years old)
3 (NBF)	New Brunswick	French	Guillaume—grade 7 (11–12 years old)
			Anne-Marie—grade 7 (11–12 years old)
			Chloé—grade 8 (12–13 years old)
			Danielle—grade 8 (12–13 years old)

[1] Many teachers taught different grades. The grade indicated in the table is the one in which they chose to video-record their lessons.

[2] Due to schedule conflicts, Carine was absent during the second phase of the data collection where the participants watched and discussed videos of other regions.

FIGURE 4.1. General Description of the Participants

The first group of teachers (Group 1) (see Figure 4.1) consisted initially of four teachers teaching in anglophone schools in Quebec (QCA). However, one teacher left the project for medical reasons. The second group of teachers (Group 2) consisted of four teachers teaching in francophone schools in Quebec (QCF). The third group of teachers (Group 3) was comprised of four teachers teaching in francophone schools in New Brunswick (NBF). For more information about the groups, please consult Chapter 1. For the Quebec francophone teacher group, we invited the school district pedagogical consultant in mathematics to participate in the focus groups. We did not consider his contributions in our analysis. However, his contributions in the focus groups sparked a lot of discussion among the teachers.

CONCEPTUAL FRAMEWORK

In analyzing our data, we needed to create a common lens to develop a collective understanding of *pedagogy*. We rooted the pedagogy in teaching mathematics; therefore, *mathematics* as a discipline to be learned was considered as a lens. Manuel and Savard's backgrounds are from French *didactics,* while Pinard, Merovitz, and Chandrasekhar's backgrounds are based in general pedagogy from the anglophone literature. Lampert et al.'s (2013) *ambitious pedagogy* was a theory familiar to all members of the research team. As will be described below, ambitious teaching practices consist of a series of practices based on principles about teaching and learning that guide teachers as they work with students. We provide further detail about the lenses used for analysis in the following sections.

Pedagogy

As noted in the introduction, our use of the term *pedagogy* is derived from the French didactics tradition and differs from the way pedagogy is used in the overall framing of the project. In our framework, pedagogy provides another lens through which to examine teachers' practices and underlying assumptions about teaching and learning. It focuses on managing social interactions, and therefore, is in the affective domain. The actions made by teachers to create a safe learning environment are considered to be pedagogy (Jonnaert & Vander Borght, 1999). All the classroom management and the materials used to engage and motivate students are thus considered to be part of pedagogy. For instance, having a poster focused on the classroom's rules of conduct or on work habits is considered to be an explicit pedagogical contract (DeBlois, 2010).

Pedagogy and didactics are two different lenses or theoretical frameworks used to study the teaching and learning process. They are both used in classrooms and influence each other. On occasion, it might be difficult to differentiate between pedagogy and didactics; this includes certain elements such as questioning and using information technology (IT). If the teachers ask questions to motivate students, it might be seen as pedagogy. However, if the question asked is about justifying a solution, it might be seen as didactics because it relates to the content being taught. The same is true for IT; using IT in order for students to be motivated by the teacher showing an animated video on an interactive whiteboard before an activity is considered pedagogy, while using IT to compare different fractions is considered didactics. We argue that differentiating between pedagogy and didactics is beneficial because it supports teachers in knowing how to intervene with students. For instance, if a student having difficulty shows frustration after a teacher proposes a task, the teacher must determine whether this source of frustration is related to learning or behavioral problems and then intervene accordingly.

Mathematics

Mathematics is another lens we focus on. This plays a key role in our analysis, as it relates to the content to be learned. We define mathematics as the concepts and processes that belong to the field. These include definitions of the concepts, their properties, representations, and symbols. Mathematical concepts were created by humans throughout history and present their own epistemology.

Mathematical concepts and processes are an important element in our framework because they play a key role in teachers' decision-making processes. Part of the work of teaching is selecting the specific mathematical ideas from history and implementing them in a learning situation (Brousseau, 1998). The teacher must also decide which pedagogical and didactic practices to use while implementing activities.

Didactics

We defined our didactics lens as the milieu created by the teacher to foster the interactions between students and the content, i.e., mathematics. All the work done by teachers to create the milieu, such as covering mathematical concepts and processes presented, using tasks and resources to support the learning process, and making demonstrations, is designed to make the content accessible for students (Jonnaert & Vander Borght, 1999, Kansanen, 2009). The teacher creates a learning environment for students so they can interact with the content to be learned. The actions made by the teachers are purposely focusing on promoting cognitive interactions between learners and the content, and therefore, we consider that these practices relate to the cognitive domain, as they focus on knowledge.

Ambitious Teaching Practices

Ambitious teaching practices are another lens we focus on as part of our conceptual framework. Ambitious teaching practices involve a set of specific actions teachers should take to orchestrate a co-constructive dialogue between teacher and students. This type of dialogue focuses on eliciting students' reasoning and responding appropriately:

> In the back-and-forth routine dialogue among students and teacher that occurs in these routine kinds of interactions, the work of the teacher is to deliberately maintain focus and coherence as key mathematical concepts get "explained" in a way that is co-constructed rather than produced by the teacher alone. (Lampert, Beasley, Ghousseini, Kazemi, & Franke, 2010, p. 131)

Ambitious teaching practices provide guidelines as to how to conduct a dialogue (Loewenberg Ball & Forzani, 2009). More specifically, ambitious teaching practices are founded on the following principles regarding teaching and learning. These principles are:

- Children are sense-makers;
- Ambitious instruction requires clear instructional goals;
- Teachers must know their students as individuals and as learners;
- Teachers must design instruction and a learning environment that supports all children to do rigorous academic work in school and to have equitable access to learning; and
- Teachers must be responsive to the requirements of the school environment. At the same time, teachers should consider why schools function as they do and how schools might improve. (Lampert et al., 2013)

Based on these principles, practices for ambitious teaching are considered to be high-quality teaching practices. These practices are the following:

- Teaching towards an instructional goal;
- Eliciting and responding to student thinking by pressing and revoicing students' ideas;
- Orienting students to each other's ideas and to the mathematical goal;
- Positioning students competently;
- Establishing and maintaining expectations for student participation;
- Representing mathematical concepts and strategies/reasoning; and
- Using a public record of student thinking.

The preceding list of practices is what we define as ambitious teaching practices. Having instructional goals in mind is central to conducting a whole class discussion (Lampert, 2001). Knowing when to revoice students' answers as a teacher, or asking other students to restate an idea to link mathematical ideas, is important in supporting students. Along with revoicing, orienting is another important practice that builds on a student's thinking in order to construct new mathematical ideas. We consider ambitious teaching practices as key elements in the work of teaching because they are the actions that support students' engagement and participation in dialogue and the learning process. Ambitious teaching practices also relate to pedagogy and didactics, depending on the teacher's intentions. For instance, orchestrating a whole class discussion is a teaching practice. If the teacher decides to facilitate a whole class discussion because students are more engaged, then this practice would be pedagogical. However, if the intention is for students to come up with an economical definition of a mathematical term, then this practice is didactical.

The model in Figure 4.2 shows how we situated our lenses when we first began the data analysis phase. It represents our vision of mathematics, pedagogy, didac-

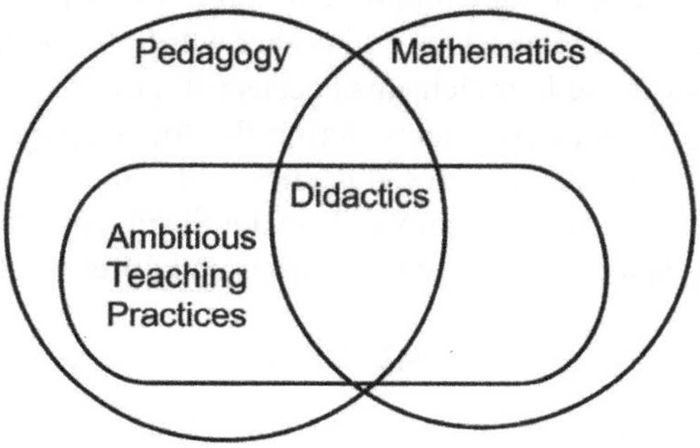

FIGURE 4.2. Model Describing the Relationships Between Pedagogy, Didactics, Mathematics, and Ambitious Teaching Practices

tics, and ambitious teaching practices. We argue that pedagogy and mathematics should be considered complimentary domains, and didactics borrowed from the French literature (Jonnaert & Vander Borght, 1999) is the intersection between them. Pedagogy represents the social, thus affective interactions or relationships between the teacher and students and among students to create a learning environment; didactics takes into account the relationships between the learners and the content, including transforming the content to be taught in a way that makes it accessible for learners (Jonnaert & Vander Borght, 1999; Kansanen, 2009). Moreover, the ambitious teaching practices are subsets of pedagogy and didactics, depending on whether they relate to highlighting mathematical ideas or to supporting the relationship between teachers and students in the classroom. Take, for example, a teacher who wants to use classroom discussion as a teaching approach. On the one hand, this teacher might use this approach to create a safe learning environment in which all students will be able to participate. This would be an example of pedagogy because it is centered on the relationships between the teacher and the student. On the other hand, the teacher might use this approach to have students debate ideas and come up with a definition of a mathematical concept or term. This second example would be considered as didactics because the teacher's intentions relate to learning mathematics.

In this chapter, we refer to teachers' representations. A representation is personal knowledge mobilized to a given situation (Brun & Conne, 1990). Thus, a representation is a personal image of a situation. For example, one person's representation about a country might be different from another person's. When thinking about Italy, someone might think about views of the sea, while another might think about vineyards. Every individual has different experiences, so their knowledge about a situation is also different. We were interested in teachers' representations about teaching and learning mathematics. Our guiding question was, "What are the participating teacher's representations of teaching and learning mathematics?" To respond to that question, we looked at what teachers said about what they appreciated while watching videos of lessons, what they said about what concerned them, what they proposed as suggestions for teaching, and what they explicitly related to themselves or, in other words, what they said they do or do not do. We chose those four elements because all groups of teachers spontaneously made those kinds of comments during the focus groups. We coded these comments as pedagogy, mathematics, didactics, and ambitious teaching practices, our lenses for analysis. We also compared the dominant lenses from each group of teachers who participated to discern similarities and differences.

METHOD

As mentioned in Chapter 1, the data obtained from the Observing Teachers project were the teacher focus group discussions. We conducted a series of focus group discussions with three groups of teachers: anglophone teachers from Quebec (Group 1); francophone teachers from Quebec (Group 2); and francophone teachers from New Brunswick (Group 3). Each group's series of focus group interviews lasted three to four full days (between 24 to 32 hours) for the first phase of the project (auto-ethnography) and two full days (16 hours) for the second phase (ethno-ethnography). During those focus group interviews, each group of teachers discussed the lessons they video recorded during the first phase of the project, as well as those selected by the participating teachers from the other regions in the second phase of the project. All focus group interviews were video recorded.

We transcribed the content of each of the focus groups in the language in which it was conducted. Research assistants coded the transcripts using the four lenses that we defined in our conceptual framework: pedagogy, mathematics, didactics, and ambitious teaching practices. We used the software Atlas.ti for the coding. Prior to the coding, the researcher responsible for the three regions and the research assistants met and made a list of sub-themes that could belong to each of the four lenses. For instance, in pedagogy, we noted the following as sub-themes: classroom management, the relationship between the teacher and students, classroom management interventions, and organization. Once a first draft of the list of sub-themes was created we shared it in Google Docs (word processing software), and the research assistants started coding the transcripts, keeping in mind that elements could be added or modified in the list. We used Google Docs so that all members of the team could see the modifications that others made to the list. When needed, the team discussed the modifications. Each transcript was coded by at least one research assistant, with some transcripts being coded by two assistants. Due to the vast amount of data, it was not possible to have every transcript coded by two people. As much as possible, we gave the assistants transcripts that they had transcribed and in a language in which they were comfortable (not all were comfortable with French). Prior to the coding, each research assistant met with the main researcher (Savard), and together they coded a short transcript to ensure reliability.

Although the participating teachers in each group freely discussed the contents of the lessons, we noticed that they tended to follow a similar structure as they watched their videos and the videos from other regions. They discussed elements that they appreciated in each lesson and elements that concerned them. They also proposed suggestions for the teacher if he or she had been present at the focus group. In addition, they discussed elements that they, as teachers, could either relate to or not. The teachers mentioned things that they also do or do not do when they teach. For instance, one teacher said, "I saw that. I do that too with my students" (Kayla, QCA). We were inspired by these types of reactions, and we focused our analysis on all the quotations that referred to elements that teachers noticed and discussed.

With Atlas.ti we made a list of common words, both in French and in English, used by the teachers while discussing elements that they liked, showed concern with, proposed, and what they related to (or not) in order to identify the quotations for our analysis. For example, "I liked it when...," "I suggest that...," "I am not too sure about…," and so on. We created Smart Codes with Atlas.ti to locate and code certain words in the quotations throughout the transcripts. For example, we created a Smart Code named "like" to capture all parts of discussion that were relevant to teachers liking what they observed. Such Smart Codes were used as an entry point to distinguish the different lenses employed by the teachers. For each group of teachers, the automatic coding function in Atlas.ti gave us a list of all quotations that contained words from the list of terms we added to Smart Code. This step permitted us to organize the quotations according to our four major lenses: pedagogy, mathematics, didactics, and ambitious teaching practices. We also organized according to the elements that teachers said that they have concerns with, elements that they proposed, elements that they appreciated, and elements to which they related. To illustrate the different lenses, we provide samples of coded quotations in Figure 4.3.

A table (see Figure 4.3) was created for each group of teachers. Once completed, we conducted a horizontal analysis of the table. We thus looked at which of the four major reactions (likes, concerns, proposed, identified with) was evident during their discussions. For example, we looked at elements that teachers liked about pedagogy, mathematics, didactics, and ambitious teaching practices. In each cell of the table, we created categories for all the elements. We determined the most discussed theme for each type of reaction (likes, concerns, proposed, and identified with). This first level analysis, by reactions, was used to provide a window into the different reactions that the teachers had. In fact, it provided us with elements to which teachers were more sensitive in regard to teaching and learning mathematics, and those that they associated with themselves (or not) by observing others teach.

We then conducted a second level of analysis, this time a vertical analysis. For this analysis, we looked at the quotations for each of our four lenses, and we looked at what the teachers discussed. We focused on the categories created in each cell for this part of the analysis process. This allowed us to determine the representation of the teachers in each group for each lens we anticipated.

We then conducted a third level of analysis comparing the three groups of teachers and the discussion themes and categories that emerged as dominant during our analysis. Below, in the results section we present the two levels of analysis for each teacher group as well as the third level analysis, which compares across teacher groups. In the presentation of our findings, we emphasize the analysis of the first level through the findings of the second level analysis. Our decision to emphasize the second level of analysis was made to reduce repetition and focus primarily on the lenses that the teachers brought to the focus groups.

The process of analyzing the data highlighted the fact that in some cases, data seemed to fit into two categories. We discuss this point after the presentation of the results.

	Pedagogy (affective domain)	Mathematics (discipline learned)	Didactics (cognitive domain)	Ambitious Teaching Practices (orchestrating whole-class discussion)
Elements that teachers liked			"She kept reminding them to come up with a generalization, always helping them refocus on the goal of the lesson, what is it they needed to do at the end of the lesson." (Camilla, QCA)	
Elements where teachers showed concerns		"Something that bothered me a lot: her equation starts with g, when she makes her visual representation, she ends with x…" (Irène, QBF)		
Elements that teachers proposed				"In this lesson, it was clear that the teacher knew where she was going with this. But there could be more room for discussion, giving more space for students to present their strategies instead of her being in the center of all the action." (Irène, QCF)
Elements with which teachers identified	"I liked the beginning, the return on previous knowledge. I liked his questioning. I saw myself a lot during this part, the questioning." (Chloé, NB)			

FIGURE 4.3. Coding Matrix

RESULTS

Different Teacher Lenses in Each Teacher Group

In the following sections, we elaborate on what was observed from the tables that we created. In the analysis, we first viewed across the kinds of quotations: what the teachers liked, what they criticized, what they suggested, and what they identified themselves with. This part of the analysis was a preliminary overview that allowed us to characterize the lens or perceptions that the teachers had as they observed each other's practices. The key focus of this chapter is the different lenses, such as pedagogy, mathematics, didactics, and ambitious teaching practices.

Here we describe the results of each group of teachers. For the first level of analysis, we present our results for each group by theme (pedagogy, didactics, mathematics, and ambitious teaching practices) presented in order based on the dominance of the themes in the data for each group. This ordering by theme dominance reveals the weight of the themes for that particular group. We will look across the groups later in our discussion.

Group 1: Quebec Anglophone Teachers

Pedagogy. For these teachers, pedagogy was about guiding and supporting students doing mathematics. Their model of pedagogy was more student-centered than traditional. These teachers believed that teachers should find different ways to teach a concept and engage students. Kayla (QCA) noted: "He uses the blackboard and chalk and yet he had all the students engaged in this exploratory and very rich content. It was student centered." These teachers emphasized that teachers need to pay attention to students with learning disabilities and provide enough examples and information to support these students. Further, they felt that it is important to write examples and information on the board, but that teachers should not guide students too much. For them, a teacher's role is to provide positive reinforcement. These teachers wanted all students to participate; they wanted them engaged with rich content. The teachers encouraged students to discuss ideas with classmates before sharing answers, and to work in collaboration with other students while the teacher walked around. A common practice is for students to present their work to other students and compare solutions. The main purpose of using IT tools in the classroom is to support learning by enhancing the content. These teachers are concerned with time constraints and would like to make the most of the time they have with their students. Tamara (QCA) was inspired by a lesson from Ontario, and mentioned these elements:

> You can tell he does this on a regular basis because they buy it. They are engaged in what they do and they are not afraid to take risks because he will validate regardless of what they were doing. It's done in a way that's entertaining and funny. He has a really good rapport with the kids and a comfortable rapport with the students. That's what sets the stage for learning. The ability to be safe and not think about what other people think, do not think about making mistakes, just explore. Look at the different ways and everything was validated and nothing was put aside.

Didactics. For the three anglophone teachers from Quebec, didactics represented teaching mathematical concepts in a structured and logical sequence while engaging students in the learning process and finding what they described as "teachable moments" to highlight mathematical ideas and make connections with other mathematical concepts, processes, and real-life contexts. They emphasized the teacher-student-content relationship. In fact, while observing lessons, these teachers emphasized some moments where conceptual ideas in mathematics were seen and others that they felt were missing. They also stressed the importance of taking time to teach concepts that build off other concepts. For instance, one participant suggested that teachers should take time teaching concepts, such as integers and fractions before introducing rates and ratios.

One aspect that these teachers looked for while observing lessons was whether the expectations were clear, and they expressed concerns when the expectations were not clear. These teachers felt a need to understand what teachers intended with their lessons. For example, while observing a lesson, a teacher appreciated the fact that she understood where the lesson was going, mentioning, "She kept reminding them to come up with a generalization, always helping them refocus on the goal of the lesson, what is it they needed to do at the end of the lesson" (Camilla, QCA). Monitoring students' activity was also mentioned as an important aspect while teaching with mathematical problems.

The group of anglophone teachers from Quebec also put a lot of emphasis on discussions and explanations of the mathematical concepts and ideas. They noted that this could be challenging because the directions where ideas may lead are not always obvious. Yet, they considered this part of the instruction mandatory, and appreciated when this was done well. The teachers also suggested that discourse should take into account as many solutions as possible and not just one.

For discourse to be effective, the four anglophone teachers from Quebec argued that many important elements must be present. A first element is teacher questioning. These teachers argued that the questions asked should invite students to reflect and explain their thinking. The questions should not only focus on procedures but also on why the procedures work. The teachers were concerned when the why questions were absent and the teacher would explain the student's answer instead. A second important element is the connections made during the exchanges. These connections include using different representations to highlight mathematical ideas, comparing different methods to solve a problem, and making links between mathematical concepts and real-life situations. Kayla (QCA) summarized these connections while observing a lesson from Ontario:

> The fact that he could entertain five different solutions of the same thing and to [be] able to want to explore each of those options with the students. The one thing that really struck me is how well he extends questions. Whether it was the question on the board or whether it is oral questions, he said, "OK yeah, what about this?" That, to me modelled curiosity. You can see in the whole class how curious and how happy they were to learn. It wasn't just chalk and talk, this is the way. Yes, he was marking

> things on the board with the chalk but the lesson was so multidimensional. I think it doesn't have to be fancy to have wonderful, wonderful learning.

Another aspect that the teachers valued in the discussion was dealing with mistakes. These teachers suggested that teachers should not deal with students' mistakes with a simple "no, that is incorrect," and move on to another student. In fact, Camilla (QCA) mentioned that:

> [A] teacher forgets a step on purpose and asks students to correct the work. It is so important for teachers to make those common mistakes so that students can avoid them in the future and also realize that their mistakes are common.

Another teacher, Kayla (QCA), talks about mistakes being teachable moments, saying:

> Every mistake is a teaching moment and that's something that needs to be reinforced in every class because and I tell my students because if it wasn't for your mistake, I will not be able to see where you are coming from. So, every mistake you make is helping me be a better teacher. So, we celebrate mistakes in this class and it's really important that we understand that we all are going to make mistakes and that's how we get better. So that, OK, what can we fix here, what needs to improve. It is a process. You want to always maintain their dignity in that process and that's what she did.

A final aspect that was also raised by the group of anglophone teachers from Quebec was the synthesis of the lessons. While observing videos, the teacher recognized the importance and the value of concluding activities by coming back to the mathematical ideas part of the lesson, but also mentioned that it was not part of their routine. The time for each class is seen as a possible obstacle. While watching a lesson from a Quebec francophone teacher, one teacher, Camilla (QCA), summarized this, mentioning that:

> I am jealous of that. We rarely get to do meaningful wrap ups if you spend that much time in introduction and then you do your activity. Before you know it, the bell is ringing. Often our wrap ups are the beginning of our next lesson. Because in 50 minutes by the time they are seated and what not, you are really looking at 40 minutes of quality teaching time. So … the 50 minutes gets shrunk and being able to pair them into groups, having a meaningful discussion, having an actual introduction, actual rap up, in a 50-minute lesson, we don't have those liberties.

This quotation shows that concluding a lesson with a synthesis of the concept learned is not something teachers do on a regular basis, but they recognize the benefits of it.

Another didactic relationship that these teachers discussed is the student-content relationship, mainly the use of resources to support learning. According to these teachers, mathematics instruction can be done while using various resources, such as manipulatives and IT. Although they believe in using manipulatives, they also argue that they should not always rely on resources, such as a calculator, for mathematics learning. The teachers were very critical of the textbooks used. They argued that, although they may be written by experts, the textbook problems do not always relate to students' lives, the level of the content presented is not always appropriate for the age group of the students, and the sequence of the chapters is not always presented in a way that favors the learning of mathematical concepts (some concepts need to be learned before others) or the links between mathematical concepts. Another resource that these teachers were critical of was the use of technology. These teachers did not seem to be influenced by resources that appear to be popular, and often mentioned the fact that no technology was used in some lessons they watched. Camilla (QCA) discusses resources:

> I think that is the underlying thing. Me as a teacher though has not changed because of technology. I think that is the difference. I am not being swayed by influences. I know what my job is when I walk through that classroom door. I know what I want to accomplish, I know my goals, I am used to picking the tools. It's the same as choosing one textbook over the other. I know what my tools will get me to meet my goals. Just because the teacher came in and said, "I have the fanciest textbook for you to use." Yeah, prove it to me. Prove to me why. Why is what you are bringing me, going to be better than what I have.

In addition, though these teachers associate themselves with other teachers who use technology, they still stress the importance of that resource to support the learning of mathematical ideas and not simply because students are attracted to technology.

Mathematics. The anglophone teacher group from Quebec believed in mathematics instruction in which students are engaged with a task or a problem to solve with appropriate resources. These teachers even appreciated when tasks focused on real-life contexts, and when the tasks went beyond the curriculum. They showed concerns when the problems given did not relate to students' experiences. For instance, these teachers talked about problems related to driving when teaching ratios. They mentioned that they use such problems because they are common problems on provincial evaluations, but they do not appreciate them as much because students do not drive at that age.

These teachers tended to analyze the mathematics involved in the task and proposed that the data used in the problem should not make the answer to the problem obvious. However, they argued that teachers should not simply give out a problem for students to solve. They should spend time, work the problem, and engage students in the problem. They were concerned when this was not done. While watching another lesson from Quebec, Kayla (QCA) mentioned:

> When he introduced the problem or when he read the problem [out] loud and proceeded to explain the problem, something was lost there. You know… I found that there wasn't enough participation from the student side, I am thinking if he were to pair them up and say OK do the problem. First read it, give them time. Especially if

> they are enriched. If they are enriched, you should always allow them to come to the conclusion before actually going and explaining on the board.

In summary, the teachers' representations about mathematics were diverse. They expressed that mathematics should be understood and not only applied, because the mathematical concepts and processes should be worked in many ways. It was important to them that students arrive at a deep understanding rather than just applying or doing. In fact, varying the strategies and spending time on mathematics was very important for them. Communicating rigorously using proper mathematical vocabulary, including using the units and different notations, highlighted their representations about mathematics as a scientific domain.

Ambitious Teaching Practices. The teachers' representations about ambitious teaching practices were focused on having students discuss mathematics. Therefore, the questions asked to students, the answers provided by students, and the class environment were important points to consider for these teachers. Thus, these teachers thought that students should explain their thinking first before the teacher's explanations, even if the thinking is inaccurate. For these teachers, having students talk about their thinking seems important; they want students to explain how they got their answers. To be effective, these teachers argued that teachers have to be patient and not interrupt the students so that students have time to revise concepts through their explanations. The teacher must also be an active listener to motivate the students. One teacher, Tamara (QCA), summarized these characteristics, saying about a lesson:

> What really struck me about the whole lesson was what an active listener their teacher was, she was constantly asking for more information and clarifying what they were asking and challenging them to go further and I thought that to me is such an exceptional quality as a teacher.

Another ambitious teaching practice mentioned by the teachers was to be conscious of the learning goal. The teacher must guide students in a positive manner and be able to guide them towards the goal, especially when students get off track. One teacher noticed, "when her students were getting off track, she guided them back by dropping clues, as opposed to being negative" (Tamara, QCA).

According to these teachers, students should be invited to present their work to other students so that all students can see the work. Then, the questions asked to students should be on how and why and teachers should support this: "Also, when he gets the correct answer from several students, he still asks many others what they got to make sure most of the students are consolidating" (Kayla, QCA). The final answer is not the most important thing to consider. The different ways to get there are valuable: "Prompting students so they can come up with their own conclusions" (Kayla, QCA).

These teachers suggest that teachers should, "look at the different ways and everything was validated and nothing was put aside" (Kayla, QCA). Thus, all students were positioned competently. Students' mistakes were considered as learning opportunities and teachers should question students on them.

Group 2: Quebec Francophone Teachers

Didactics. For the Quebec francophone focus group, didactics are about the teacher-student relationships, the student-content relationships, and the teacher-student-content relationships. The teacher-student relationship is about supporting students when they are presenting their work. It is presenting a way to communicate their work to other students.

The student-content relationships are about providing support to students to interact with the content. Such relationships focus on providing students with resources for learning such as manipulatives, graph paper, posters on the walls, and formulas written on the blackboard. The resources also included processes such as taking notes or work methods. Student work is also a resource when it is used to start a discussion and then build on.

The teacher-student-content relationships are about the teachers' support for the student-content relationship, such as providing explanations, examples or demonstrations, and questioning students. The format or the structure of the lesson is also an important point for them. When providing explanations, teachers should not tell the students too much in order to keep high expectations. In this sense, doing a partial demonstration on the board might be preferable; the teachers did not think it was necessary to show the whole demonstration in order to have students understand. Thus, not explaining or unpacking the process at the beginning of the lesson is important for having students look for deeper knowledge. It is important that explanations provided by teachers are accurate and avoid creating learning obstacles. The explanations could also be seen as processes to perform the work. In this case, writing the steps on the board after the lesson is a way to focus on the student-content relationships. The selected examples by the teachers should take into consideration presenting representations before doing the work. Using students' answers to provoke deeper reflection could be seen as using this work as an example, or to probe the thinking further through questioning. In fact, having students present and discuss their work in front of the classroom is a strong way to deepen the interactions between students and content. For these teachers, questioning students is about providing accessible questions. Questions should pace the learning process and not create gaps when making sense of the concepts. For these teachers, questioning students using different representations is a way to help students make different connections. Using students' questions in discussion is also important. Johanne was particularly appreciative of a teacher she observed in one lesson because she frequently questioned students and supported a mistake that a student gave. "Lots of questioning, and she [the teacher] even worked with the mistake. She wrote the wrong answer on the board and questioned the student about it so he could realize his mistake" (Johanne, QCF).

These teachers' representations regarding the format or the structure of the lesson framed the interactions between the students and the content. They felt the lesson should reactivate students' previous knowledge, followed by note taking. The aim is to support students in a co-construction process, so that it was important to provide enough time for them to think and work. The task presented to students was expected to be rich and relevant; it is better when the context presented is linked to daily life. The teachers felt that tasks should help students understand the mathematics, and therefore, the representations coming from the task should be linked to the manipulation of material (manipulatives) and the equation. Teachers felt that the lesson synthesis presented at the end was an important time to allow students to think and reflect on the content. Apparently, most of those characteristics were those they found in exemplary lessons. Carine (QCF) argued that in secondary 1, students have already learned most of the concepts taught in elementary schools, thus it would be a great opportunity to create more exemplary lessons:

> You know most of the content we cover in Secondary 1 has been learned in Grades 5 and 6. This could be a great opportunity to give students more interesting activities where they [students] could explore concepts with manipulatives, you know make it fun for students, you know when they can get out of the classroom after and go like "wow that was fun!" Even though students have learned most of the content we cover before, doesn't mean that we have to teach it (traditionally).

Pedagogy. For these teachers, pedagogy is about relationships and well-being in the classroom and having teaching tools. Pedagogy involves having an environment that positively supports students. Components might include strong classroom management and good relationships with students. Having a participative classroom where students are engaged was also valued. The teachers were aware of not creating too much stress for students by not having them work in isolation at the black board. The teachers saw the role of the teacher as a person who is active, dynamic, and who circulates to monitor student thinking while students are working. As for IT tools in the classrooms, these teachers mentioned tools that they found useful: a document camera to show students' work; a blog for sharing ideas; and a digital document created for the Interactive board (SmartBoard). To them, those tools are engaging for students and helpful for the teachers. Roger (QCF) synthesized most of those characteristics while observing a lesson from New Brunswick:

> You know it's really, you see with all the things she (teacher) had. She used the interactive board, the poster with the formulas. It was not static, it was dynamic. She walked around in class smiling. The students were sitting at their seat, but it was still dynamic and with all the support she had and gave, I found it was great.

Ambitious Teaching Practices. These teachers' representations of ambitious teaching practices highlighted the role of clear didactic intentions while teaching.

The teachers discussed the idea that having clear didactic intentions allows teachers to better guide students in discussions, and thus, provide opportunities to question students with open questions about their thinking, their conceptions, and their strategies. Questioning students is also used when comparing different students' strategies. Teachers suggested that all students' answers should be discussed at some point. Revisiting content when needed in the discussion is also important. Leaving enough time for students to think in a discussion is also necessary. These teachers felt that the teacher should not correct students during a student discussion; it is better to guide students' thinking to make them realize the mistake. Asking students to reformulate something important is a way to help students understand the content. Irène (QCF) summarized those aspects when commenting on a lesson by a teacher from New Brunswick:

> In this lesson, it was clear that the teacher knew where she was going with this. But there could be more room for discussion, giving more space for students to present their strategies instead of her being in the center of all the action.

Having clear didactic intentions allows teachers to identify expectations; therefore, expectations should be communicated to students before having students do the work.

Mathematics. The representations that these teachers have about mathematics are more specific to mathematics. For them, mathematics is not only a field containing concepts and processes. It is also a field that requires deep thinking and reasoning skills, as well as rigor. Therefore, mathematics is about making conjectures, generalizations, and proofs. For instance, while discussing a lesson on probability, Roger (QCF) said:

> You know, students will rely on their impressions. If they do the experiment 10 times and they find something close to what they believe will happen, they will conclude this. This is why we should create experiments in class as much as possible for students to manipulate and work with these ideas.

In addition, for these teachers, mathematics is also a field that highlights finding different strategies to solve problems, and using manipulatives and modelling to explore mathematical ideas. For instance, these teachers discussed different concepts and their meaning, the mathematical representations of those concepts, and the vocabulary used. Carine (QCF) had firm beliefs in proposing tasks where she can discuss strategies with her students, as she explains:

> For me, I find it important and I do it with my students, discuss problems and different ways to solve the problem. For instance, with mental math, how we can make this calculation. It is always fun when my students come up with two or three different ways to solve a problem.

This group of teachers was observant of the mathematical ideas that were presented on the board by the teachers they observed in the videos. The teachers would make comments when ideas were incorrect, for instance, when they observed a teacher teaching a lesson on equations and he or she used a different variable midway through the problem.

Group 3: New Brunswick Francophone Teachers

Pedagogy. When it comes to pedagogy, a principal element that seems to strongly represent the four francophone teachers from New Brunswick is the relationship between the teacher and students. These teachers seem to value an environment where every student, including students with special needs, feels a part of the learning community. It appears that they want their students to appreciate them as teachers. For instance, Guillaume (NBF) mentioned:

> Every year, I ask my students to give me feedback on me and my teaching. I even tell them not to be afraid to be critical or even negative if that is the case. If you hate me, you can tell me, I say to them. This helps me improve. And students are usually very honest with me.

In addition, the teachers feel that they have a role to play in this relationship as well. In fact, these teachers would often discuss qualities they observed in teachers in the videos, such as being dynamic in class, having a sense of humor, using humor in a good way, being interesting, and being able to keep the students engaged and interested in the lesson.

Another element that appears to be important to these teachers is the learning environment. This was shown by the fact that the teachers always mentioned how they considered their students with learning difficulties while planning lessons to find strategies to support their learning, and some teachers implemented options for the more advanced students. Although differentiation was important, these teachers also did not want to label their students. They showed concern when they saw lessons where teachers called out a student as good in mathematics. In addition, classroom management was another important element of pedagogy that these teachers seem to value. In fact, they preferred a calm learning environment in which all the students are engaged in the learning. They shared concerns when the students in a lesson from a teacher in Ontario were too loud or doing other things in the classroom instead of working on the task. "This was a great lesson, but I felt that it was a bit too noisy in the room, a bit more than I could tolerate" (Anne-Marie, NBF).

The third element related to pedagogy that represents these teachers is the learning activities implemented in classrooms. In fact, these teachers believe that various approaches should be used in teaching. They sometimes feel that group work becomes monotonous at times. Other strategies should be used, such as games, videos, and sometimes chalk and talk. One teacher often mentioned how she functions using the flipped classroom. No matter what teaching approach is used, the teachers from New Brunswick seem to argue that three elements should be present. First, the students need to be engaged, and second, technologies should be used. One teacher, Danielle (NBF) mentioned:

> Students practically eat technology for breakfast, lunch and dinner, so using technologies keeps them interested in classrooms. We have to go with what interests them. I feel that we can always use technology when we teach, even if it's just showing a short clip.

Third, visual support needed to be present in all learning activities. These teachers find that visual support is important, especially for students with learning difficulties, and shared concerns when they observed lessons that were too auditory.

Mathematics. As for mathematics, one element that was important for these francophone teachers from New Brunswick was mathematical vocabulary. These teachers believe that vocabulary should be emphasized throughout the lesson. They argue that mathematical terms should be defined at the beginning of the lesson, and they should review the terms at the beginning of each lesson to activate prior knowledge. They showed some concerns when this was only seen at the end of lessons. The teachers also believe that students should use proper mathematical vocabulary while sharing ideas with others. The teachers from New Brunswick appreciated when they observed teachers who insisted that their students use correct mathematical terms when explaining their thinking.

The second element that is important to the teachers from New Brunswick is the representation of mathematical ideas. We observed different points of view about the value of using different representations of mathematical concepts and processes. For instance, one teacher proposed that they should show the method for solving linear equations symbolically and with algebra tiles at the same time for students to be able to grasp the concept. However, in another case, teachers felt that using a traditional method, for instance the rule of three for solving proportional problems, makes things easier.

Teachers giving "tricks" or shortcuts to students to help them remember or memorize mathematical concepts and processes was also part of their representation of teaching. The teachers from New Brunswick admitted to giving students tricks. However, they did argue that tricks are not enough. Students should still know the mathematical ideas behind those tricks or shortcuts.

Didactics. As for didactics, the four important elements for the francophone teachers from New Brunswick are the learning sequence, the mathematical tasks used in lessons, the resources to support the learning, and the discourse between the teacher and the students. The teachers from New Brunswick tended to put an emphasis on teaching mathematical concepts in a logical way. In fact, they appreciated seeing teachers reminding their students what they have learned, what they will learn, and what is to come. They also believed that lessons should be constructed in a logical way. For instance, in the introduction of the lesson, they believe that teachers should define concepts and activate students' prior knowledge. In addition, there should be a synthesis at the end of each lesson to review what was learned.

As for the tasks, these teachers tended to believe that good tasks or problems should be used during instruction. Teachers appreciated lessons in which they observed students working on tasks and discussing them instead of teachers lecturing students. Some even made positive comments about the choice of numbers used in a problem. They believe that real life situations should be used as much as possible. However, this was not consistent. While thinking of students with learning difficulties, they felt that some activities might be too difficult for them. For instance, having a task where students create an equation for their peers to solve was possibly "too difficult for my students with learning difficulties. I cannot imagine them being able to do such a task" (Anne-Marie, NBF).

As for the resources to support learning, the teachers argued that visual support should be used as much as possible. They encouraged the use of different manipulatives and resources, and technologies were possibly at the top of their list. They even proposed that physical manipulatives be used instead of drawing whenever possible. For instance, Chloé proposed that real measuring cups could be used to show fractions instead of drawing them on the board. However, they also thought that some manipulatives presented a challenge for learning certain concepts.

As for the interactions between teacher and students, these teachers from New Brunswick believe in taking the time to clearly teach mathematical concepts and processes. This could be done by asking open-ended questions so students can explain their thinking, and teachers can respond in ways that enhance student comprehension. Learning mathematics for these teachers requires students to be able to apply appropriate procedures and understand the ideas behind the concept. Anne-Marie (NBF) summarized these facts by mentioning:

> It was so great how the teacher took the time to explain the concept to the students. It is long, but it is very important too. We need to take the time for this if we want our students to learn mathematics, and not skip steps or assume that they all understand, even the ones with learning difficulties. It was really impressive to see that teacher do that.

Chloé agreed. "I take the time myself, and sometimes I am behind compared to my other colleagues who teach mathematics. We follow the same planning. Even if I fall behind, I don't regret it. I see positive results after."

Ambitious Teaching Practices. In terms of ambitious teaching practices, the francophone New Brunswick teachers value a lesson where open-ended questions are used with students. They suggested prompting students with questions that include how they solve a problem and why this strategy works. These teachers felt concerned when they observed teachers question students about their strategies and then would summarize the students' answers with an explanation of why it was correct. "I think teachers should ask students why this works also, not just how you solve it," said Anne-Marie.

In addition to the practice of questioning students, the teachers from New Brunswick also found it important to involve students as much as possible to avoid a "ping-pong match" between the teacher and student. These teachers believe that other students should participate in the discussion. For instance, other students can be asked if they agree or disagree and explain why. Students could also be polled and asked who else got the same answer or solved a problem the same way. These teachers also believed in making sure that all students can contribute to the discussion. Anne-Marie (NBF) mentioned:

> I am guilty of doing this at times. I ask a question and I immediately ask the first student who raises his hand to answer. However, lately, I am waiting for more hands to come up and will ask students who raise their hand later to answer. Sometimes, the students look at me and wonder, "What are you waiting for?" But it is important that we give time for students to think. Of course, our stronger students will often have an answer first, but the others can do the same if we give them time. And they need to participate as well. They are in the class too, just like the others.

Analysis of Different Lenses across Teacher Groups

In this section, we compare the findings across the three groups of teachers according to our lenses: pedagogy, didactics, mathematics, and ambitious teaching practices.

Pedagogy. In terms of pedagogy, all groups showed some similarities among their representations about classroom management. The teachers emphasized the need for creating a safe and supportive learning environment for all students, including students with learning difficulties. They also shared the idea that students should be exposed to various types of activities, and they agreed that visual support must be included in all activities to support student learning. The relationship between the teacher and students was also an important element for them. In fact, the three groups talked about the importance of having a teacher who is dynamic, has a sense of humor, and is able to maintain student engagement in mathematical activities. We felt that the teachers from New Brunswick may value this more because they included different types of students, such as those with learning difficulties, in their discussions. In addition, as stated previously, Guillaume never hesitated to ask for feedback on his teaching from students at the end of the school year because he wanted to improve each year. Although all three groups seem to feel as though all students should be involved in the activities, the group of anglophone teachers from Quebec and the New Brunswick group of teachers talked more explicitly about students with learning difficulties.

Teaching resources to support learning is something important that all three groups discussed, with IT resources receiving the most attention. The teachers from New Brunswick felt strongly about using technology in activities. They found that technologies should be included, as they keep students engaged. The anglophone teachers from Quebec were more critical about technology and thought that technology should not be used simply because it "looks cool." They suggested that technology should be included if it supports learning. These teachers often applauded when they observed teachers engaging students in rich mathematical content without the use of technology. As for the francophone teachers from Quebec, they seemed to agree with both the New Brunswick teachers and the anglophone teachers from Quebec. For them, IT should be included in activities to engage the students and to give them visual support for their learning.

Didactics. For didactics, we identified similarities and differences in how the groups view the teacher-student relationship. The group of anglophone teachers from Quebec and the teachers from New Brunswick discussed the importance of all their students being active in the learning activities. The anglophone teachers from Quebec also mentioned that activities must encourage students to come up with their own conclusions. However, a 50-minute class period time limitation was an issue, as they felt that the amount of time was quite short for interacting with students. The teachers from New Brunswick also discussed the importance of taking the time needed to explain mathematical concepts through open-ended questioning so students can understand the how and the why behind the concepts. Conversation from the francophone teacher group from Quebec seemed to be different from the other two groups. These teachers talked about the importance for students to come up with clear solutions so other students can understand them.

As for the relationship between the student and the content, all three groups talked about using a variety of resources, including IT, to support students' learning. Although all teachers talked about the importance of access to multiple resources, the importance of note taking seemed most important to the group of teachers from New Brunswick. The anglophone teachers from Quebec suggested that IT should not be the only option to consider.

Another aspect from the discussion relating to the student-content relationship is the value of teaching mathematics using real-life contexts where students can relate the mathematics to their everyday lives. This element came mostly from the anglophone teachers from Quebec and the teachers from New Brunswick. The anglophone teachers from Quebec also added that the task can go beyond the curriculum. However, taking into consideration students with learning difficulties, the teachers from New Brunswick seem to question this idea for some of the tasks they observed in lessons.

All groups discussed the teacher-student-content relationship, specifically with a view to how to structure a lesson. For the four francophone teachers from Quebec, a lesson should reactivate students' previous knowledge, followed by note taking. They also highlighted the importance of using questions that are accessible to students to delve deeper into reasoning. More specifically, the teacher should not unpack the process at the beginning but should write the steps on the board after the lesson, which would be part of a synthesis. The Quebec francophone group was also very specific about what to do during a lesson. They had a set format for a lesson: the teacher should provide accurate explanations, examples/demonstrations, and ask questions to students without providing too much information. In addition, for these teachers, a teacher must avoid creating learning obstacles and present different representations before doing the work. The tasks presented to students have to be rich and relevant, linked to students' daily lives, and lessons should include manipulatives, a representation, and an equation. For them, students must have enough time to think, work, and present and discuss their work, as discussion is important.

The francophone teachers from New Brunswick had a different vision: teachers should provide a definition for mathematical terms at the beginning and not the end of the lesson. They focused on the three-phase structure of a lesson, which progresses as preparation (including activating prior knowledge and proposing the task), realization (students work), and integration (including synthesizing the content learned and applying knowledge in other contexts). The anglophone group from Quebec highlighted the importance of sequencing the content by using spiral learning in teaching mathematical concepts, something that they suggest textbooks do not offer. For them, lessons must have teachable moments with lots of connections between mathematical concepts and methods. As part of this, the teachers viewed exploring mistakes as an important opportunity for a teachable moment. They also emphasized that teachers should always have clear expectations and remind students of what they need to "come up with" (Kayla, QCA) mathematically.

Mathematics. In terms of mathematics, all three groups discussed the domain, how to do mathematics, and what to focus on when teaching mathematics. For the two groups of francophone teachers, mathematics referred to the meaning of mathematical processes and the vocabulary employed. Mathematical representations were also part of the domain for the francophone teachers from Quebec. For the group of anglophone teachers from Quebec, mathematics as a domain was represented using different notation. The groups showed differences in how to approach mathematics. The anglophone teachers from Quebec referred to varying the strategies with mathematical concepts instead of just working one way, while their four francophone Quebec colleagues referred to being rigorous and developing understanding. The francophone Quebec teachers also mentioned making conjectures and generalizations as important aspects of mathematics. The francophone teachers from New Brunswick emphasized using proper vocabulary. The terminology employed by teachers has to be relevant and used frequently. This teacher group from New Brunswick also mentioned that using different representations of mathematical concepts and processes is an important focus. In addition, they said that teachers might give "tricks" or shortcuts to students, but the focus should be on understanding. In contrast, the francophone teacher group from Quebec added that the focus should be on strategies, manipulation, and modelling. It seems that the francophone teacher group from Quebec had a deeper understanding of the mathematical activity than teachers in the other two groups.

Ambitious Teaching Practices. When comparing the findings of ambitious teaching practices, four main elements emerged: goals and intentions; discussion and questions; comparing student work; and listening and thinking time. For both groups of teachers from Quebec, goals and intentions were considered important to their practices, while they were not discussed as much by the New Brunswick teacher group. In terms of discussion and questions, all three groups acknowledged the importance of leading discussions on student solutions by asking open-ended questions. As for comparing student work, all three groups of teachers reported using this teaching practice. They agreed on the fact that discussing student answers and mistakes are great learning opportunities. In terms of listening and using time to think, all three groups of teachers mentioned these were essential for supporting students' mathematics learning. While the four francophone teachers from Quebec highlighted this for helping students realize their mistakes, their anglophone colleagues talked about the importance of being an active listener and having patience, and the four New Brunswick teachers talked about the importance of waiting for more hands to be raised before calling on students. This was essential to provide time for students to think.

DISCUSSION

As we began the data analysis, we were quite aware and accepting of the different lenses that we brought to the project. For Savard and Manuel, this lens primarily focused on didactics. For Pinard, Merovitz, and Chandrasekhar, the focus was more on pedagogy in general. Having all worked with the ambitious teaching practices, we all saw how this lens or approach could benefit teaching as well as our analysis of teaching. Coming into this project, we saw a vision for our conceptual framework. We saw didactics as an intersection that was characterized by the space where mathematics meets pedagogy. In working with the data, we realized that our model for didactics was far too simple. After working with and structuring the data, we observed that there were too many instances where data failed to fit neatly in one category or another. What followed was the realization that our lenses did not fit into just two domains with intersections representing didactics and ambitious teaching practices. This made us consider the possibility that mathematics, didactics, and pedagogy involved three different spheres, which all had the possibility to overlap. This was confirmed when we set out to map three of our four lenses in a diagram. By considering all of the possible intersections, we confirmed that all of the intersections included data from our analysis. This left us then thinking of our last lens, the ambitious teaching practices. More specifically, we were thinking of where this fits in our new conceptual framework. Fundamentally, the ambitious teaching practices could span the three lenses of mathematics, pedagogy, and didactics. Consequently, the ambitious teaching practices fit neatly in the center of our conceptual framework overlapping all of our lenses and their intersections. To see our revised version of our conceptual framework, refer to Figure 4.4.

As we analyzed the data, we questioned our initial ideas about the major similarities and differences among the different groups of teachers, especially con-

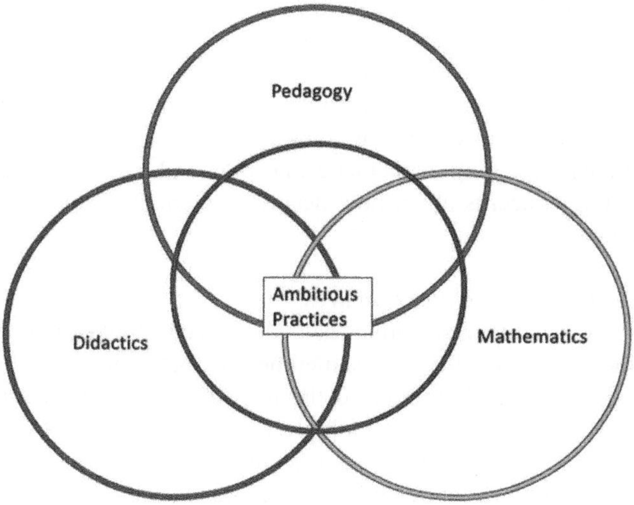

FIGURE 4.4. Revised Conceptual Framework

cerning our conceptual framework. Thus, some ideas could be interpreted into subcategories that we did not have in our original framework. We looked at our data to see which of the subcategories from our revised conceptual framework were present. Thus, we came up with these ten subcategories: six subcategories with the intersection of two domains, three subcategories with the intersection of three domains, and one subcategory with the intersection of all four domains:

Subcategories with the Intersection of Two Domains

Pedagogy/Didactics

One element that can be part of the intersection between pedagogy and didactics occurs when the teachers were talking about the tasks. All three groups talked about the importance of bringing elements from students' daily lives into their tasks. They also talked about the mathematical concepts and processes embedded in the tasks. Those elements would refer to didactics in our model. However, the three groups of teachers also mentioned that tasks should be engaging and even amusing for students. This aspect highlights the relationship that the teacher creates with the students, thus part of pedagogy according to our model.

Pedagogy/Mathematics

Some of the discussions related to the intersection between pedagogy and mathematics referred to student anxiety and how the teacher prepares students for what is to come. Many teachers shared their anxieties, wondering if they prepared the students well enough with the mathematical content for the following year, according to the various courses that students may choose, including, for instance, enriched mathematics. Some even talked about the anxiety that they and their students experience with provincial examinations. Not knowing what questions are asked on the examinations, teachers always questioned if they went deep enough with the mathematical concepts they taught. We often hear about students' anxiety towards mathematics. However, it was interesting to hear about the anxiety that teachers have with teaching mathematics. Dealing with student anxiety is part of pedagogy, while concerns about how deeply mathematical concepts are taught refer to mathematics.

Pedagogy/Ambitious Teaching Practices

The topic of visual representations had many teachers reflecting on their practices, as some commented that they sometimes do not put enough representations on the board. These teachers noticed the importance of representing students' thinking on the board, and using multiple representations, such as the number line. This element relates to ambitious teaching practices. Their reason for using this practice related to pedagogy because they concluded that many learners are visual learners. Thus, representing students' thinking would support these learners.

Didactics/Mathematics

Some groups of teachers looked closely at how the mathematics was presented during the lessons. They made comments about some aspects related to mathematics that teachers should avoid during teaching to help students avoid certain learning obstacles. Some examples of issues related to solving linear equations include inconsistency with the variable in an equation, always putting the variable on the left side of the equation, and putting the answer on the right. Teachers also noticed that some teachers always place the zero at the center of the number line. In these examples, teachers refer to key mathematical concepts, such as variables in equations, and whether there is a center in a number line. At the same time, the teachers made comments related to didactics. For example, they commented on how to present mathematical ideas using part of a number line according to the context of a problem.

Didactics/Ambitious Teaching Practices

An element that describes the relationship between didactics and ambitious teaching practices was found in classroom discussions. All three groups of teachers talked about the importance of using ambitious practices by asking open-ended questions. It was important for these teachers that students are able to explain how they solved problems and why the strategies work. Representing students' thinking is another practice that teachers addressed. While discussing those teaching practices, all three groups of teachers emphasized the importance of paying attention to elements that relate to didactics. For instance, they talked about not giving away too much information to their students in order to maintain the cognitive demand for a task. Another example is identifying the mathematical ideas to emphasize.

Mathematics/Ambitious Teaching Practices

As mentioned previously, the teachers acknowledged the importance of using multiple representations to signify mathematical concepts and ideas, for instance, encouraging drawings and symbols to represent fractions. They also valued supporting students to be able to make links between the representations. Some groups of teachers also argued for the importance of using standard mathematical representations. For instance, some questioned the use of different notation such as using a colon in representing ratios.

Subcategories with the Intersection of Three Domains

Pedagogy/Ambitious Teaching Practices/Mathematics

One element that related the components pedagogy, ambitious practices, and mathematics was teacher concern about the students with learning difficulties. Some groups of teachers focused on how mathematics was presented in lessons, and questioned whether measures were taken to support all students including those with learning difficulties. Some groups, like the francophone teachers from New Brunswick, wondered if students with learning difficulties would be able to explain or unpack the mathematical ideas of some mathematical tasks that they observed in lessons, and whether there was enough visual support to promote the learning of all students. Some groups, like the anglophone teachers from Quebec, looked at how the questions asked of students and the visual support enabled students to grasp the mathematical content. These conversations focused on how ambitious practices helped to highlight the mathematics involved to support students with different needs, highlighting the affective aspects of learning.

Pedagogy/Ambitious Teaching Practices/Didactics

When the three groups of teachers talked about the discourse they have with students, elements from pedagogy, ambitious practices, and didactics were present. For instance, in relation to ambitious practices, all groups of teachers talked about the importance of prompting student thinking by having students explain their thinking and justify their explanations. These teachers talked about the importance of asking open-ended questions: the *how* and *why* questions. In addition, most teachers talked about the importance of representing students' thinking. The visual support seemed to be an important aspect to teaching for all three groups. In relation to didactics, the teachers talked about the importance of being an active listener. Furthermore, they highlighted the importance of noticing students' conceptions and in finding the "teachable moment" from a student's thinking. In relation to pedagogy, the teachers talked about different elements that should be important to keep all students engaged during discussions and for creating a safe learning environment. For instance, teachers from New Brunswick mentioned the importance of not always calling on the first student who raises his hand. The anglophone teachers from Quebec talked about having patience with students when they are explaining their ideas.

Didactics/Ambitious Teaching Practices/Mathematics

All three groups of teachers talked about taking the time to teach mathematical concepts in a structured and logical way. These discussions involved making sure that the mathematics was presented with appropriate notations, as well as making sure that didactic contracts (Brousseau, 1998), such as changing the variable or always putting the variable on the left side of the equations, were respected. These discussions also included using practices, such as building on students' thinking and reasoning to highlight the mathematical ideas involved, or in some cases mathematical misconceptions (as some valued mistakes). For instance, on many occasions, teachers would comment about how some ambitious practices would help highlight mathematical ideas and create a didactic learning condition. For instance, Quebec francophone teachers noticed that using a rectangular model of fractions instead of a circular one (ambitious practice of representing a mathematical idea) to represent the quantity one half and one third (mathematical ideas) permitted students to see why one half was greater than one third (didactic).

Subcategory with the Intersection of Four Domains

Mathematics/Didactics/Ambitious Teaching Practices/Pedagogy

Having clear didactic intentions allows teachers to better guide students during discussions, and thus, to find opportunities to question students with open questions about their thinking. These types of questions probe students about their conceptions, strategies, and reasoning. Thus, when creating learning conditions for students and tasks for students, a teacher takes into consideration the mathematics to be learned, student reactions to the task, the inclusion of a relevant context, and ways to implement the task such as which questions to ask. From our perspective, this is the most important aspect of teaching, and the most complex aspect to attain.

Further Conceptualizing Didactics

When analyzing the data, we attempted to situate the lens under which the teachers' conversations occurred. By depicting the teacher, the subject matter (mathematics), and the student in a triangle, we were better able to visualize the interactions between them. The student interacts with the teacher and this constitutes pedagogy. The teacher interacts with the mathematical knowledge to support student learning by creating the best learning conditions possible and by making the mathematical knowledge accessible to the student. As for didactics, this is what links the student to the teacher and to the mathematical knowledge. For didactics to occur, the teacher must make subject-specific pedagogical decisions based on his/her understanding of the knowledge. As such, didactics can be seen as a bidirectional arrow that extends from the top of the triangle (the teacher) to the middle of the base (students in interaction with the mathematical knowledge) as seen in Figure 4.5. The bidirectional arrow means that the teacher interacts vertically with the students in interaction with the mathematical knowledge, and the feedback from this vertical interaction informs the teacher. This didactics arrow is the core of the teacher's work, which entails: 1) creating a learning environment that best suits all students in order to learn mathematics. This implies that not only the physical environment (material, resources, room, and peer interactions) is taken into consideration, but also 2) working conceptually on the knowledge to be learned by students to make knowledge accessible for all of them. This a priori analysis (Brousseau, 1998) of the knowledge to be presented to students goes beyond looking at the tasks: it is an analysis of the concepts themselves and the epistemological obstacles that they might have in relation to students' thinking.

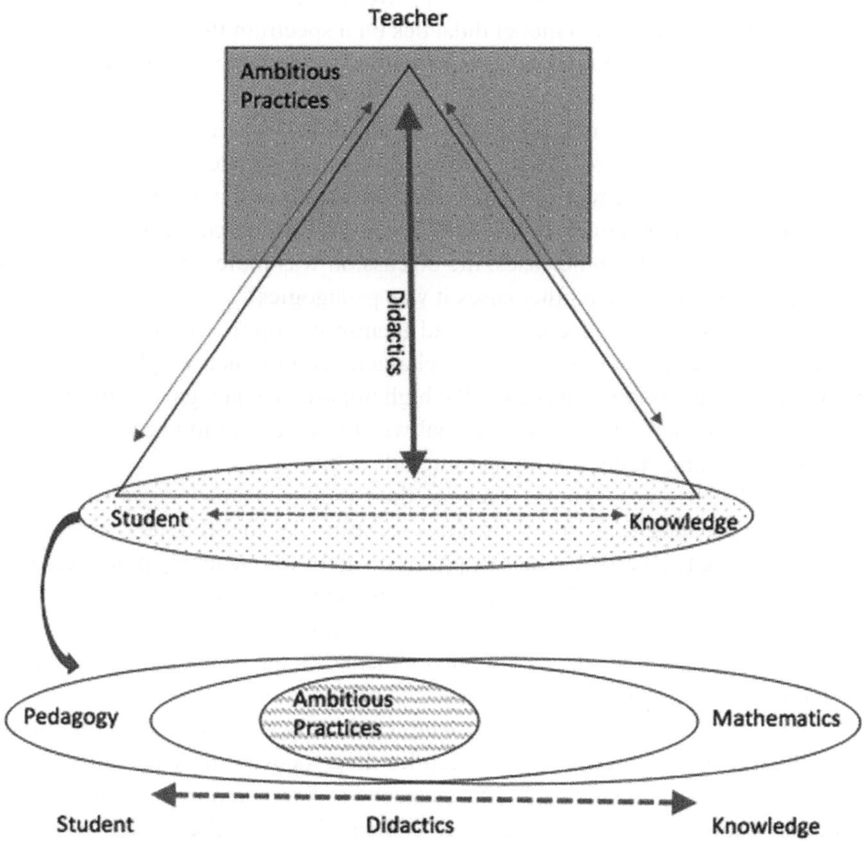

FIGURE 4.5. Model to Conceptualize Didactics

In attempting to model this, we again asked ourselves where our ambitious teaching practices were situated. In the interactions with students, the ambitious practices constituted specific decisions that the teacher undertook to engage students in dialogue and to promote powerful discussions about mathematics. As a result, the ambitious teaching practices were centered on the teachers, their interactions with the students, and their interactions with knowledge. The ambitious teaching practices thus include the top portion of the triangle working down towards the base, as seen in Figure 4.5.

As we moved forward in the data analysis, we found that classifying teacher discussions under the lens of didactics did not adequately characterize the discussions that we were observing. In trying to summarize our findings, we expected to find some teachers talking more about didactics, with others focusing more on pedagogy, mathematics, or ambitious teaching practices.

In trying to situate these different discussions, we observed that most of the teachers in the Quebec francophone group were talking about didactics most of the time. This allowed us to model didactics on a spectrum that connects the student to knowledge or, in our case, mathematics. In the bottom of Figure 4.5, we can see the arrow that connects students to knowledge as part of a cross-section from our revised conceptual framework. In our model, we can consider didactics as a spectrum that moves between knowledge and the student. This connection can vary as it flows between pedagogy and mathematics and can even include ambitious teaching practices. It is in this zone that we observed the majority of our teacher discussions. In some cases, the discussion was more on the mathematics end of the spectrum and in other cases it was pedagogical.

The New Brunswick teachers situated themselves on the pedagogical end of the didactic spectrum. This was mostly characterized by their emphasis on technology, on student relationships, and the high importance they gave to differentiation in the classroom. Their primary goal was to make learning accessible to all students, including students with learning difficulties.

For the Quebec anglophone teachers, their practices and focus also situated them on the pedagogy end of our didactics spectrum. What differentiated the Quebec anglophone teachers was their emphasis on ambitious teaching practices. This was demonstrated by their focus on using questioning, student questions, and student thinking as central to effective class discussions. They believed that it was beneficial to let the students explain the mathematics themselves and to continue explaining it during the lesson.

The Quebec francophone teachers were situated near the mathematics end of the didactics spectrum. This can be seen by their focus on how the students interacted with the mathematics. For these teachers, it was important to be efficient with the sequencing of the mathematics content. These teachers were not afraid to push past the curriculum, but their focus was on teachable moments that made connections between the mathematical concepts and the methods. This conceptualization is important, because this is the first-time didactics and ambitious practices are contrasted and discussed together.

LIMITATIONS

Collecting data at many sites and in different languages was sometimes challenging. For instance, some teachers watched other teachers teaching in another language. Some assistance was needed by teachers to make sense of teaching episodes. The researcher and the research assistants helped the teachers by translating specific words. However, it is important to note that the data we collected during the focus groups were not translated; the data were transcribed in the language used in the meetings with teachers. Data were analyzed in French and English simultaneously, based on the essence of what was being communicated by the teachers. The researcher and research assistants spoke sometimes in French, sometimes in English, and sometimes using both languages during the discussion.

Having a huge amount of data and the involvement of many research assistants brought some challenges that might have impacted our results. For instance, the fact that coders sometimes had different interpretations of a code might be seen as a limitation of this study. Even if we constructed a model to develop our shared understanding, most of our codes emerged through the coding phase, and thus, could have been interpreted differently. Another limitation was the search in Atlas.ti using the smartcode "like." In French it was not a problem, because the term "j'aime" was used for describing what they like. In English, "like" was also a synonym of "as," which provided different information. We had to look at the data without the search of "like" for the anglophone group of teachers. Therefore, we found more quotations for them than for the groups of French speakers.

CONCLUDING REMARKS

In concluding our analysis, we realize that although it is helpful to consider the different lenses that researchers and teachers bring when observing each other's lessons, the distinction is not always clear. After all, pedagogy, ambitious practices, and mathematics are not mutually exclusive of didactics. Thus, didactics covers a broad spectrum as it includes, at some point, more or less pedagogy and more or less mathematics. This point is important because it is something to consider when training teachers. Having a clear focus on the role of those lenses might guide us in our work with teachers, either when doing research with them or while supporting them. This clear focus also brings other areas to investigate, such as how teacher training influences teachers' practices and discourse about teaching mathematics. When working with the teachers, all groups discussed didactics, whether focused on pedagogy or mathematics; however, as the teachers discussed, each group appeared to have a dominant discourse. We found these different discourses quite rich. We wondered about the influence of the teachers' initial teacher training: how did this initial training impact their perception of didactics and their stance as mathematics teachers?

REFERENCES

Brousseau, G. (1998). *Théorie des situations didactiques* [*Theory of didactical situations in mathematics, 1970–1990*]. Grenoble, France: La Pensée Sauvage.

Brun, J., & Conne, F. (1990). Analyses didactiques de protocoles d'observation du déroulement de situations [Didactical analysis of observation protocols of the unfolding of situations]. *Education et Recherches*, *90*(3), 261–285.

DeBlois, L. (2010). Et si on pensait les troubles du comportement autrement? [What if behavioral problems were thought of otherwise?]. *Nouvelles CSQ Bulletin du CRI RES*, *23*, 21–24.

Jonnaert, P., & Vander Borght, C. (1999). *Créer des conditions d'apprentissage: Un cadre de référence socioconstructiviste pour une formation didactique des enseignants* [*Create learning conditions: A socioconstructivist frame of reference for didactic training of teachers*]. Bruxelles, Belgium: De Boeck Supérieur.

Kansanen, P. (2009). Subject-matter didactics as a central knowledge base for teachers, or should it be called pedagogical content knowledge? *Pedagogy, Culture & Society*, *17*(1), 29–39.

Lampert, M. (2001). *Teaching problems and the problems of teaching*. New Haven, CT: Yale University Press.

Lampert, M., Beasley, H., Ghousseini, H., Kazemi, E., & Franke, M. (2010). Using designed instructional activities to enable novices to manage ambitious mathematics teaching. In M. K. Stein & L. Kucan (Eds.), *Instructional explanations in the disciplines* (pp. 129–141). New York, NY: Springer.

Lampert, M., Franke, M. L., Kazemi, E., Ghousseini, H., Turrou, A. C., Beasley, H., . . . & Crowe, K. (2013). Keeping it complex: Using rehearsals to support novice teacher learning of ambitious teaching. *Journal of Teacher Education*, *64*(3), 226–243.

Loewenberg Ball, D., & Forzani, F. M. (2009). The work of teaching and the challenge for teacher education. *Journal of Teacher Education*, *60*(5), 497–511.

CHAPTER 5

DISCUSSING FRACTIONS

Teacher Conversations about their Teaching Practices

Sophie Pinard, Dominic Manuel, and Annie Savard

This chapter presents an exploratory analysis on how three groups of teachers perceive their own and other's practices while actively teaching fractions in the classroom. The different ways to represent a fraction and how those representations support students' understanding were discussed among the three groups of teachers. The analysis of these discussions provides insights into what is important for the teachers in relation to teaching and learning fractions. It also points to the different knowledges that teachers lean on during the teaching of fractions.

This chapter is related to Chapter 4, in which we explored the representations of three groups of teachers: three Quebec anglophone teachers, four Quebec francophone teachers, and four francophone teachers from New Brunswick. In this chapter, we used the same discussions of the same three groups, but we place our emphasis on one particular topic: how teachers perceive their own and other's practices while actively teaching fractions in the classroom. We chose this topic because all teachers were asked to video record an introductory lesson on a topic related to fractions. This permitted us to look more deeply into the participating teachers' discussion about a common mathematical topic, one that is known to be challenging for students to learn. Although we used the same data as used in

Researching Pedagogy and Practice with Canadian Mathematics Teachers, pages 93–115.

Chapter 4, here we used a different method of coding to enable us explore teachers' perceptions of their practices.

Prior to this study, the research team at McGill University (Merovitz, Pinard, Chandrasekar, Manuel, and Savard) conducted a preliminary study on the representations that these teachers used and how they taught fractions in their classrooms. The team presented a poster entitled *I can do that too! Teachers' discussing how they represent fractions* at the *40th Conference of the International Group for the Psychology of Mathematics Education (PME)* in Hungary in 2016. Results showed that the teachers used various representations, such as models (both rectangular and circular), a number line, cooking instruments, and sets of objects, and that they argued that the practice of representing fractions was a way to support student learning. We continue from that preliminary study, and delve deeper into the data by focusing on the teachers' perceptions of teaching fractions in general. The goal was to determine how the participating teachers from the three groups view teaching and learning fractions, in all aspects of the work of teaching. What is important to them? What do they put more and less emphasis on when teaching fractions? What do they find easy and challenging? What are their obstacles when teaching fractions? We intend to bring some answers to these questions and others. Our guiding question was: how do the participating teachers from Quebec (anglophone and francophone) and New Brunswick view or perceive teaching fractions?

In this chapter, we first present the framework, developed by Shulman (1986), which outlines three types of teacher knowledge: subject matter content knowledge, pedagogical content knowledge, and curriculum knowledge. We use this as our conceptual framework in our analysis of the teacher's discussion on fractions. Second, we present a brief discussion about fractions. In that section, we define fractions, present multiple interpretations of fractions, and acknowledge information about teaching fractions from previous studies. Third, we present our data collection method. We describe how we coded the transcribed discussions from the three focus groups to extract all the quotations in which the participating teachers talked about fractions. Fourth, we present and discuss the results. The results are presented in three main subsections that reflect the three types of knowledge defined by Shulman (1986). Finally, we conclude this chapter by briefly summarizing the results of this study.

A CONCEPTUAL FRAMEWORK FOR TEACHER KNOWLEDGE

Trying to understand how and why teachers teach the way they do is a topic that has been an "active field for a long time" (Da Ponte & Chapman, 2006, p. 461). Although some models that exist attempt to explain the work of teaching, more work needs to be done to capture in full detail the work that teachers do in their classrooms with students (Elbaz, 2018). One model that attempted to identify the types of content knowledge that teachers have as practitioners is Schulman's (1986) framework. He proposed three categories of teacher knowledge: subject matter content knowledge, pedagogical content knowledge, and curriculum knowledge. As part of subject matter content knowledge, teachers must understand the content they are teaching. This understanding should go beyond knowing how to apply the mathematics and include a comprehensive, global, and connected conceptual understanding of the subject matter. Extending past facts and concepts, part of this type of understanding requires teachers to understand "why" mathematics is so.

Pedagogical content knowledge revolves around the teacher's ability to utilize powerful representations of the concepts being taught. This includes "analogies, illustrations, examples, explanations, and demonstrations" (Shulman, 1986, p. 9). Also in this category is an understanding of what makes certain concepts or ideas difficult for students. This includes knowledge of student preconceptions, conceptions, and misconceptions.

The last category, curriculum knowledge, involves an understanding of the educational program, the instructional materials available, and the methods of assessment. We used these types of content knowledge to frame our analysis of the discussions about teacher practices that are explored in this chapter. Further, because the practices discussed in this chapter focus on the teaching of fractions, it follows that these categories of content knowledge tie specifically to a conceptual understanding of rational numbers.

A BRIEF DISCUSSION ABOUT FRACTIONS

A fraction is a number that can be expressed in the form p/q, where p and q are integers and q is not equal to zero. The fact that a fraction can be written using two numbers is somewhat misleading to those students who conceptualize both numbers independently, rather than considering p/q in its entirety as a number (Rosar, Van Nieuwenhoven, & Jonnaert, 2001).

Fractions can have many interpretations. Chapin and Johnson (2006) defined five interpretations of fractions as described in Figure 5.1.

Fractions are considered a complex mathematical concept for students to grasp, and many questions regarding how teachers can facilitate such an understanding are difficult and researchers still continue to bring specific answers to those questions (Behr, Harel, Post, & Lesh, 1992). According to Behr et al. (1992), from a pedagogical view, fractions take on numerous "personalities" that can be considered from two perspectives: the mathematics of quantity (Schwartz, 1988); and as an element of a multiplicative conceptual field (Vergnaud, 1983, 1988). In a similar vein, Kieren (1993) elaborated on how rational numbers are part of a quotient field, which adds complexity for understanding. A simple example of this is the connection between addition and multiplication. Normally multiplication is considered as a repeated addition that makes elements larger; however, with rational numbers, multiplication is independent of addition and can be "seen best as a composite function where the numbers themselves become a multiplicative operator" (Kieren, 1993, p. 55). In addition, in the realm of a quotient field, the

A fraction can be:		Definition	Example(s)	Representation
1a)	A part of a whole	The idea that the total or "everything" can be divided into portions or equal parts.	One of four equal sections of a circle. Or 5/8 of a pie.	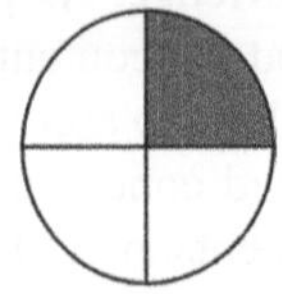
1b)	A part of a set	A certain quantity of a total of identical objects. May be discrete when objects are not to be sectioned.	2 out of 5 star stickers. Or 2/3 of 12 identical candies.	
2)	The result of dividing two numbers or a quotient	The idea that to divide the collective we need to partition it into a desired number of equal-sized parts. Can be based on the idea of equal sharing or the action of separating every item into the same number of parts. Can use a repeated division.	2 chocolate bars divided equally amongst 8 people. 3 loaves of bread divided equally amongst 9 people.	
3)	A ratio of two quantities	The comparison of a part to a part or a part to a whole.	2 apples for every 3 bananas. 2 apples: 3 bananas. Or 4 girls for every boy (part to part) or 16 boys for each class of 24 (part to whole).	2:3 
4)	An operator	The idea that an object can be magnified or shrunk using a fixed proportion. The idea of stretching and shrinking to scale.	A small sized replica that is a scale version of an original image. Or zooming in or out with a microscope.	
5)	A measure	The idea that each line segment can be broken into smaller parts. The idea that we can re-partition a part.	A ribbon cut in half or in fourths. Or drawing 1/4 by locating 1/2 of the 1/2. Or lengths on a number line, using units (cm) and subunits (mm).	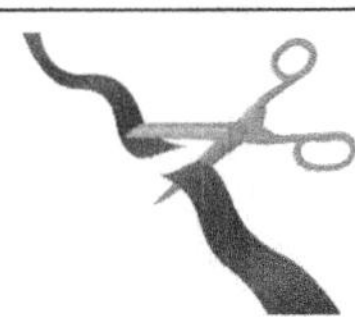

FIGURE 5.1. Different Interpretations of Fractions

number 1 plays a dual role of being both a dividable unit and the foundation for a multiplicative inverse.

Although the conversations discussed in this chapter relate to fractions, the goal is not to study the teaching nor the learning of fractions themselves. This is something that has already been studied, notably by Ball (1993), Confrey, Maloney, Nguyen, Mojica, and Myers (2009), Kieren (1993), and Lamon (2012). Interestingly, although the teaching of fractions in classrooms has often been the subject of extensive research, little has been published about how teachers perceive their own practices and the practices of others while actively teaching fractions in the classroom. This is why we concentrate here on the teachers' comments about how fractions are taught in classrooms. We suggest that focus group conversations, where teachers comment on their own and others' teaching, offer opportunities to better understand how teachers view teaching and learning fractions. It is through observations and discussions regarding what others have done, have changed, and have omitted that a window into what teachers consider important is possible. The essence of what teachers see as important can be witnessed through what they understand other teachers are doing and applying in the classroom.

METHOD

The following analysis is based on the data that uniquely relates to the discussions regarding teaching and learning fractions during regional focus group sessions with three anglophone teachers from Quebec, four francophone teachers from Quebec, and four francophone teachers from New Brunswick. We used the transcripts specifically from these three groups, as we were the research team responsible for gathering data with them.

Our first step was to extract the transcripts that referred to teaching and learning fractions. To reduce the data on the conversations targeted for this investigation, we used two methods. First, we applied a coding process using Atlas.ti to extract and code for the key word *fraction*. Once we completed that first level of coding, we located the quotations in the transcripts that contained the word *fraction* and copied them into a document. Second, we reviewed the sections of the teacher discussions that included the selection and discussion of teachers viewing the video on fractions. This step enabled us to extract other quotations in those discussions that related to teaching fractions but did not contain the word *fraction*. We also looked for quotations that related to fractions but did not include the key word *fraction* in the other discussions (when teachers discussed typical or exemplary lessons) to assure that we collected all the pertinent quotations for the study. Turns of talk were selected to be included as part of this data. On occasion, this selection extended past a simple turn of talk to include the following turn of talk when the idea or sentence was continued, to include the full meaning of the contribution to the discussion. These steps were conducted for the three focus groups. We used pseudonyms (presented in Chapter 1) to identify the participants.

With the selection narrowed to the specific quotations related to teaching fractions, we created an initial file for analysis. We first took a holistic approach by reading the data as a totality in an effort to establish a direction for a coding scheme. Then, we selected quotations in the text based on their contribution to ideas regarding fractions, and the way they were observed to be taught. We reduced these relevant sections of the text to a code that conveyed the key thought behind the teachers' contributions. Next, we used those keywords to compile a list of key ideas. We then reviewed the words on this list and reduced them to new larger and encompassing ideas that connected different contributions in the text to each other. We called those larger ideas *themes*. Finally, from this reduced list, themes were established that connected into categories of big ideas using Shulman's (1986) three types of content knowledge (subject matter content knowledge, pedagogical content knowledge, and curriculum knowledge). That is, we attempted to classify the categories and themes according to that framework. We next present these themes that emerged from the data, and structure them within the appropriate form of knowledge.

FINDINGS AND DISCUSSION

For this analysis, we structured the teacher contributions according to Schulman's (1986) three forms of knowledge. Within these three types, there is also an overall structure that we established, based on several rounds of coding. In the subsections, we elaborate on the findings for each of these categories organized within the knowledge types. It is important to note, however, that not all teachers' comments fit neatly into one category. Some statements are included in one category although they could have fit into another. In these cases, for simplicity, we included them in the section where we thought they seemed to best reflect a particular category of knowledge.

Subject Matter Content Knowledge on the Meaning of Fractions

As discussed earlier (see Figure 5.1), one complex facet of fractions is that they can have different *interpretations*: parts to wholes or parts of sets; the result of dividing two numbers; a ratio of two quantities; as operators; and as measures (Behr, Harel, Post, & Lesh, 1992; Behr, Lesh, Post, & Silver, 1983; Kieren, 1993; Lamon, 2012). Some of the discussions the three groups of teachers had about fractions could be connected to all different interpretations of fractions. More interestingly, not all of the five interpretations of fractions came up in the discussions in all groups of teachers.

One way that these groups of teachers discussed their expression of fractions, in terms of the *part to whole relation*, was through the use of circles and rectangles as models. With much of the emphasis on the representations about dividing circles and rectangles into parts, it would be reasonable to expect many of the discussions to be about the part to whole relationship. Although there were many occasions where the discussion was about parts and wholes, there was no explicit statement made by any teacher about the part to whole interpretation of fractions. This created an interesting question for the research team that might merit more investigation: are the part to whole concepts of fractions so present in teaching practices that they are no longer apparent to educators? How can we be more explicit about this practice? It may be that the part to whole relationship seemed evident to them. With this set of data it was hard to answer this question.

Along with part to whole concepts of fractions are *part to set* interpretations. An interesting discussion came about during the teacher exchange regarding fractions in terms of parts of sets among the Quebec francophone teachers. Roger talked about how it is important to move away from the idea of a fraction as a collection of objects. He noted that it is a representation that is used a lot in elementary school and that students should "stop there" (Roger, QCF) and move on to other interpretations of fractions. He explained that the reason why he was concerned had to do with the operation of addition in this context. In these types of exercises, students are asked to add sets together. He gave the example of having five boys out of six students in one group and then three boys out of four students in another group. By combining the groups, you end up with eight boys out of ten students. This provides a misleading understanding of how to add fractions as it is evident that five sixths plus three fourths is not equal to eight tenths. He explained that this sort of reasoning prevents students from thinking about how to determine the common denominator that is necessary to add the two fractions correctly.

Through the selection process of lessons on fractions, the New Brunswick teachers chose a video where the multiplication of fractions was demonstrated with an *area model* (Figure 5.2), which can be seen in interpretations 1a and 1b from Figure 5.1. The New Brunswick teachers were not the only group that reacted positively to this video; other groups discussed the idea of multiplication with the area model. Camilla and Carine from the Quebec anglophone group also reacted. For Camilla, the use of the visual representation of fractions helped make the connection between the multiplication of whole numbers in general and the multiplication of fractions. She stated, "And using visual representation in order for them to see, to visualize, that, how they are the same and different" (Camilla, QCA). For her, it seems that the knowledge acquired about multiplication can be extended to a new set: rational numbers. However, there can be an element that prevents this knowledge from transferring over; from what Camilla said, this barrier can be eliminated through visualization. Approaching subject matter content, Carine also contributed, "I really liked the part with the passing from visual representation to the symbolic representation, which is good to help the understanding of students" (Carine, QCF). These comments show the beginning of a common trend focusing on the importance of visual representation, which will be discussed under the category of subject matter content.

Another interpretation of fractions discussed by the groups of teachers was the *measurement concept of fraction* (see Figure 5.1, interpretation 5). Although

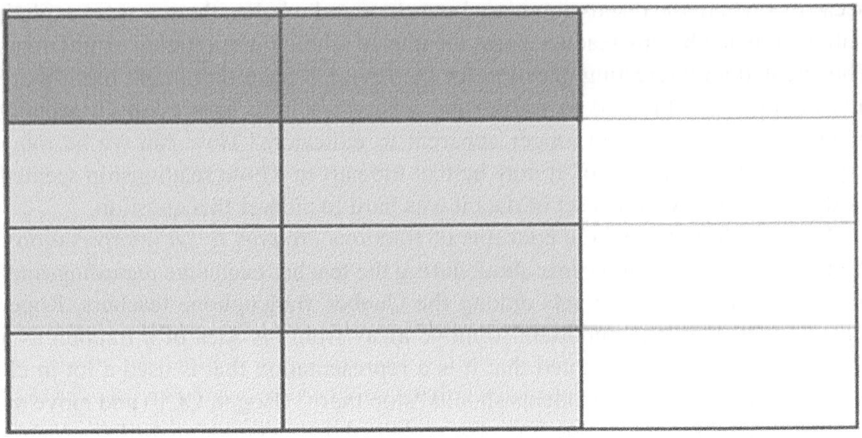

FIGURE 5.2. Area model representation of a multiplication of one fourth by two thirds.

these types of discussions could also be considered as subject matter content, we chose to include them here as they express teachers' understandings of fractions as numbers on the number line. These discussions came up several times in the Quebec francophone group. Carine commented on how she liked that another teacher took the time to place fractions on the number line and took the extra step of finding another fraction that is between the two. She followed up by stating that she would not have necessarily thought of doing that. Roger spoke about how he is incorporating the number line in his own practice. He shared, "I used the number line this year. Like to represent two fifths, I did a rectangle above zero to one and I divided it in five and we identified it on the number line" (Roger, QCF). The teachers agreed that number lines were underutilized even though they were important to help students better integrate the concept of a fraction. As an example, Carine proposed that, in their selection of a video to share with the other teacher groups, they should select what Johanne did because it incorporated a number line in the lesson. Figure 5.3 is an example of the number line teachers discussed. Although in their discussion, the teachers agreed that number lines were important, there remained the question of how to incorporate them. Johanne chose to use a number line as her first representation; this was something that the other teachers wanted to understand better. They did not seem convinced by the idea that the number line could be used before any other representation.

Meanwhile, in the same group of teachers, the discussion of denominators brought them to the notion of a common denominator. The idea here was to think about the different multiples for each number, looking across to determine one that is common. For this, Irène liked another teacher's choice of a fraction over four and a fraction over six. She stated, "And it is wonderful, to arrive just at

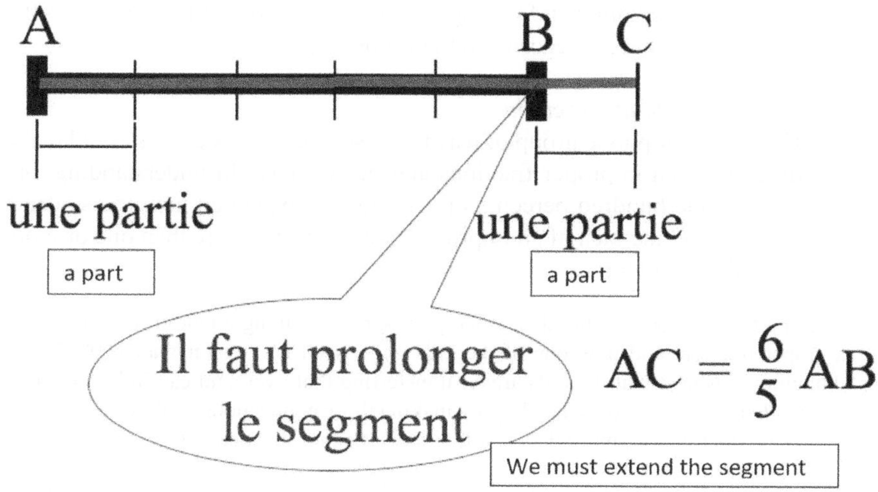

FIGURE 5.3. A number line representation of fractions taken by a teacher from resources that come with the interactive board. (Translations in English are in rectangular text boxes.)

working the common denominator, you went a little further with another trick. It is really from six and you are going from four, you went and looked for all the multiples" (Irène, QCF).

The discussions among the teachers also extended past the idea of a number in the fraction form of *a*/*b* and extended to the use of decimal and percent equivalents. For example, the discussion of decimals was something the teachers from New Brunswick considered. They saw the decimal form as a powerful and simpler way to compare fractions. As part of this discussion, they accepted that students could use a calculator to convert fractions to decimals as long as they were capable of explaining why a certain decimal corresponds to the fraction. They noted that this was the method of choice for their students, as explained by Chloé (NBF):

> I asked them to place the fractions in increasing or decreasing order. … I have students that would have automatically changed them into decimals. They understand, "Oh, it is the same thing," that is a lot easier. Why rack my brain with a common denominator?

Pedagogical Content Knowledge

Three major categories emerged from the discussions related to pedagogical content knowledge: student thinking about fractions; representing fractions and their importance in learning; and linking fractions to everyday life. We discuss the findings for each of these categories in the following.

Student Thinking about Fractions

The Quebec anglophone group of teachers discussed how students could make connections between improper fractions and percentages. In understanding that the whole was one hundred percent, Tamara (QCA) explained how the students were able to extend the idea of an improper fraction being more than one or more than one hundred percent:

> And the fact that one of the kids, when you were explaining about how to define a proper fraction, and how one of the kids goes "well it's greater than 100%." I thought it was neat that the kids are … transferring that over, and can link it to so many other things. It shows that they are making those connections … they're thinking, okay, fractions are more than just fractions. It's pies, it's percentage.

Another mathematical concept discussed by the same group was the idea of *equivalency*. Kayla remarked on how one of the other teacher's videos really touched on the importance of equivalency for her. "It looked like they were doing equivalent fractions and there wasn't necessarily a talk about equivalence" (Kayla, QCA). In the Quebec francophone group, Roger and Irène emphasized the importance of visuals for demonstrating equivalencies. They discussed the need to reinforce diagrams when connecting equivalent fractions, such as three fourths and six eighths. For them, it was important to demonstrate and separate the pieces using examples such as "cakes." It was about emphasizing where the multiplication by two halves came from. Irène explained that to do this she shows the representation and then divides each piece in two, which implies having double the smaller "pieces." Irène (QCF) also returned to the idea of preferring rectangles when showing the representation of three fourths and six eighths, saying,

> And why I like better, say the representation as a rectangle, it is that I will like separate my whole, I just separated it in two…So then if I keep the same whole, and I just separated it right down the middle. There, I see that three fourths is six eighths.

The last mathematical topic touched upon by the groups of teachers during their conversations was the concept of *operating on fractions*. This was especially the case for division, the operation that the four Quebec francophone teachers discussed as being especially challenging for students. Carine (QCF) commented:

> But I find that for the division of fractions, the verbalization of all that, it is really not evident for students to understand that, for example, when you divide by four well then it becomes four times smaller, but then it's four times bigger.

This brought the teachers back to iterating the importance of visuals to facilitate student understanding. Roger shared how he takes the time, as a class, to do "the drawings so that the students see the division and all." For him, the visualization was an important part of explaining to students that they need to multiply by the inverse fraction when they perform the operation of division. Roger understood the importance of representation as part of his practice and shared, "But usually, when I do division of fractions, I will do it [have the students make their own representations]. I will make the link with a recipe, and I really use the idea with cups and all that" (Roger, QCF). These comments echoed the previous emphasis teachers attributed to making visual representations discussed in the beginning of the analysis.

One important aspect that the groups of teachers talked about as being important to them was the use of appropriate mathematical vocabulary. An attention to language and terminology was present in the discussions of all three teacher groups. This was stated to be important for both teachers and students. With respect to her goals and expectations for students, Danielle from New Brunswick remarked, "It is important that all students know their vocabulary." However, it was primarily the case that teachers noted the teacher's own use of or focus on vocabulary or terminology in the classroom, as Kayla (QCA) noted:

> I loved just staying with the vocabulary; she had these vocabulary sheets, beside her smart board written in a number line [format] of what those vocabulary words meant. So, you can see that she is paying attention to math literacy, because she really wants to make sure they understand those words.

In the teachers' discussion, we can see that the mathematical vocabulary was an important element that spoke to the quality of the lesson. Sometimes it was very specific to how a particular concept was being taught. An example of this was shared by Johanne, "I really found your verbalizations excellent for your whole part comparing the numerators, comparing the denominators with the visual representations … we can't do better than that" (Johanne, QCF). Other times the comments were more global with respect to the whole lesson, as when Guillaume remarked, "And she used a lot of good vocabulary" (Guillaume, NBF). This important focus on language is something that we can understand from the teachers as being an important part of discussing mathematical concepts and processes. Still, as part of the analysis it was difficult to determine into which category these remarks fit. On the one hand, terminology is an important part of communicating mathematics, so it fits in somewhat with the mathematics of the fraction. On the other hand, terminology also expresses a need, purpose, or larger goal that the teachers have and connects somewhat to our next category.

Representing Fractions and Their Importance in Learning

Much of the conversation the teachers had was linked to their pedagogical content knowledge of fractions. These sorts of exchanges often included ideas about how to present the topic, and about the students acquiring the knowledge of "why" things are so. These types of conversations highlighted the teachers' pedagogical content knowledge as they demonstrated understandings of common student conceptions, preconceptions, and misconceptions.

One aspect discussed by the teachers was the visual aspect of fractions. Talking about the representation of fractions using images connects to pedagogical content knowledge. It is not surprising to say that much of the discussion regarding each other's practices focused on the visual aspect of teaching fractions. This was especially apparent in the New Brunswick teachers group when one of the teachers used an area model for teaching multiplication of fractions. This sparked a conversation in which the teachers discussed the importance of using visuals to teach fractions. Unfortunately, an understanding of the need to promote the visual aspect of fractions was met with a limiting factor of time, which was associated with curriculum knowledge. This can be seen in teacher comments that reiterated that time was a factor, and they wished they did not have to run through the curriculum. These teachers stated that they knew that they needed to spend more time using visual representations and did not do it enough. This is apparent in the following comment by Anne-Marie (NBF) in reaction to the visual approach to multiplying fractions:

> I like it a lot; I realize that I do not do enough visuals, that I jump always even too much to the symbolic. We just finished, because fractions were not on the program and we saw ourselves missing time and we do a lot of explicit teaching there, where this is how we do it. … It takes a lot of time when you say I want visual going to symbolic. I imagine that you do not give a lot to do because the steps are somewhat elaborate.

In the above comment, we also noted a discussion about how teaching becomes focused on how we "do" certain things. Danielle said, "I think that we should go with the visual representation before … the steps" (Danielle, NBF), which shows that the steps of doing things are important to her, but not until the concept is acquired. This importance of understanding "how" contrasts with a central goal that this group of teachers said was important to them. This goal is about having the students understanding and explaining "why" the math works. Guillaume addressed this, as well, after observing a fellow teacher using a visual representation to make a demonstration. He said that the question, "Why is it true?" (Guillaume, NBF), is important. Visual representation was powerful for the New Brunswick teachers. For them, students needed to first visualize fractions before they could learn how to apply them. In addition, the students' ability to represent fractions visually was considered an important way for the teacher to determine when students had acquired conceptual understanding. For Guillaume, the important question to answer was, "When do we say 'Okay, there, they have done fractions, they have visualized them. It is done. They can move on to what is [a] more comfortable [interpretation]'" (Guillaume, NBF).

We can see that all the teachers accepted that they need to focus on the visual representation of fractions. This is also illustrated by Roger (QCF), who discussed the lack of visuals in another teacher's video and proudly shared about the use of visuals in his teaching. He said:

> What I noticed is that it is very abstract. There are not many drawings. I use many drawings with the students.... I think that as math teachers we do not spend enough time on the drawing, on being very explicit.

Here we can see that Roger started to address an important factor that he thought contributed to the lack of comfort observed as students try to make meaning of rational numbers—the word *abstract.* At another moment in the teacher meetings, Roger continued this thought further saying, "It is just numbers, but remove the image, the abstraction part of fractions is very complex to them [students]" (Roger, QCF).

Although all groups of teachers communicated the general need for visuals, they also discussed something specific about them. One of the things that they talked about was the use of circles and rectangles while representing the fractions (as seen in Figure 5.4). This can be seen in the teacher discussions about how they thought they utilized circles too much in their representations and do not favor the rectangle enough. Kayla (QCA) explains:

> I liked it because I find we all sometimes fall in the trap as showing fractions as circles. She did circles and all of a sudden went to rectangles to explain the kid's answer. I thought it was so great because she was connecting, and we are not always doing the pizza. It's a new way to see it, looking at it as a fraction and not necessarily a circle. She is broadening their understanding.

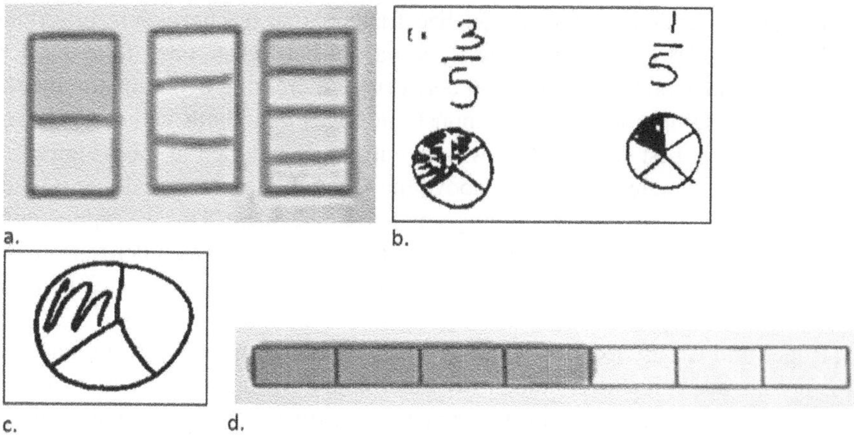

FIGURE 5.4. Examples of Representations of Fractions Used by Teachers.

The use of circles and rectangles was echoed by Guillaume when he said, "You know, like the famous pizzas, at some point, the circular forms often used, she also brought some forms of rectangles, so that there were other forms to present" (Guillaume, NBF). It can be seen from their comments that there is often a connection between fractions, pizzas, and pies. The connection with a fraction to another circular object, such as a pie, emphasizes the tendency to consider fractions as parts of a whole that is circular. Teachers' preference of using circles over rectangles can be more easily understood when we consider a comment made by Irène. Another teacher had remarked that she often used circles, and to this Irène responded, "Yes, oh why? Habit maybe, often I will, it can be traced. It sucks to say, but it is really easy to trace on the board. I could always go get other forms, some rectangles, uhm, speed maybe" (Irène, QCF). Johanne also noticed that, unlike her, another teacher from her group often chose the rectangle for many of the representations. This led her to make the realization, "Me, when I listen to you, I see that you do several representations with rectangles. And me I don't do it. Maybe it is lacking a little bit from my end" (Johanne, QCF).

These were the sorts of moments where the groups of teachers realized that they needed to have some variety in ways of teaching fractions in the classroom. This was part of an admission of how teachers could improve their representations as well as an acknowledgement that variety of representations should be part of better teacher practices. This was illustrated by Anne-Marie's appreciation of another teacher's imagery and representations. She noted, as a positive, that the teacher she observed focused as much on using circles as rectangles. Recognizing that rectangular representations were successfully implemented by another teacher shows that Anne-Marie knows it should be part of her teaching practice.

In addition to the need to integrate rectangles more frequently into their work, the three groups of teachers discussed the superiority of the rectangle. This was particularly discussed when the Quebec francophone teachers were observing each other's videos teaching fractions. As part of her lesson, one of the teachers chose to demonstrate the connection between an increasing denominator and the size of the resulting parts. This demonstration was done both with a circle and with rectangles: first separated in two, then in three, and then in four (Figure 5.4 a). With the rectangles placed one above the other, the teachers said that it made the fact that the pieces were becoming smaller more evident. This was appreciated by Carine, who remarked, "And to see it earlier, with Irène, with her representation with the rectangles [being] a lot more convincing than the circle" (Carine, QCF).

Roger and Carine took the time to discuss the concept of fraction, and that it is not always evident for students to determine how many such parts it would take to recreate the whole. This was the case for Roger as he shared that for many students this idea is hard to grasp, being either not acquired or not a "reflex" (Roger, QCF). This is something that Roger said he worked on in class, having the students filling a cup using thirds of a cup. Carine also talked about how she observed this difficulty with students. Carine (QCF) shared with the group:

> We will ask them to represent three fourths, two fifths … that goes well, then if we get to sixths, then oops, we just created an obstacle, but the principle of dividing into seven, they understand that really well … but to go to a level that is more abstract, to say how many sevenths are in a whole, then all of a sudden, we just asked a question that makes no sense.

For the teachers in New Brunswick, the use of visual representations in the classroom went beyond the teacher's practice; it was also something that they expected of their students. This can be seen when Chloé said that visual representation was important, not only as a tool for the teacher to convey the mathematics, but also as a means of assessing student understanding: "Yes. And show me using a drawing. Using imagery, what does it mean?" (Chloé, NBF). This was something with which the other teachers agreed. For these teachers, students can only resort to using "mathematical tricks" if they truly understood the mathematical concept. For this, the onus was on students to prove that they understood why a trick works. The teachers thought a good way for the students to demonstrate this was by using a visual representation. For fractions, they wanted to see proof in the form of a drawing.

Some of the discussion revolved around rational numbers in different forms. As part of this, the teachers in New Brunswick discussed their observations of the Quebec anglophone group. They enjoyed the other teacher's choice of a fraction that was a mixed number. Anne-Marie (NBF) shared:

> I liked her use of images, because it hooked me, I said, especially when she had a mixed fraction. So, when she had 4 and a half, she did not want to know where the 9 came from, she wanted to know where the 8 came from. … And there, it was making a link with the whole, and the denominator before finding the right fraction or mixed number.

She further contributed that these numbers are difficult for students to work, "Because often, they will not know with a mixed number how to divide their drawing.… When you have a mixed number and you ask [them] to make a drawing, often it can take some time" (Anne-Marie, NBF).

Another concept that appeared in the discussions was the notion of the whole, its size, and how it matters. An example of this can be seen in Guillaume's sharing of his own practices when he discussed how eating a quarter of a pizza means nothing if the actual size of the pizza is unknown. He illustrated this, sharing that the students remarked, "Yes sir, if the pizza is large like that, and we take a quarter …" (Guillaume, NBF), implying that the size of the whole matters. This big idea about fractions was also accompanied by a discussion of another important element, the idea of equipartitioning. Camilla demonstrated the complexity of student understanding involving the importance of equal shares. On one hand, she explained that students have acquired the concept that working with fractions involves the division of a whole into parts. She maintained that this is learned early on saying, "They understand why they are dividing, because it's the number of pieces that are in the whole. So they do that, because that is the process that's taught in elementary schools" (Camilla, QCA). On the other hand, although there is an understanding of dividing into parts, the parts are not always understood to be equal. Camilla shared that this is the case when the students work with thirds, "Some of these kids don't realize that you have to separate this pie into equal pieces, you know, especially with thirds. They're cutting it in half and then cutting two quarters" (Camilla, QCA). These comments seemed to indicate that there is a range of student understanding with respect to the concept of equal shares.

From this discussion, we can see that fractions can be "more than just fractions" and can be expressed in different forms. The idea of equivalency between a number expressed as a fraction, as a decimal, or as a percentage was something that Guillaume shared when he talked about an activity he likes to do in class. He talked about a game that he plays with students, where he distributes cards that represent numbers in different forms: fraction, decimal, percentage. The students have to find other cards that represent an equivalent of their card. Then he asks students to line them up by value from greatest to smallest.

The next theme referred to the goals and approach of the teacher or lesson. This mostly tied in to pedagogical content knowledge and knowledge of curriculum. As an example for this category, some amount of discussion revolved around the importance of focusing on the "why" and the "how" in the classroom. Similar to the discussion about the use of calculators to convert fractions within the New Brunswick teachers group, the Quebec anglophone group of teachers said that it was important for them to be able to explain why a "trick" or procedure works. This can also be seen in Camilla's (QCA) comment:

> If the student just says "Ah! But I just divided, and it gives me that," it is not an answer, like that...You cannot use that strategy. If you use it, that strategy, you have to prove why it gives that answer.

From Camilla's contribution, we can see that the teachers were open to students using shortcuts or tricks, as long as they were capable of proving why the strategy worked. In addition, the importance of explaining "why" was the responsibility of the student not only to quickly find an answer, but also to explain why it works. From this group of teachers' comments, they conveyed that they also felt that this was a necessary step for the teacher. For Chloé (NBF), the explanation of why something was true should be integrated with the visual:

> We forget to go back to the basics ... The importance of going back and giving them the visuals explaining *why*. Why do we multiply? The whole by the denominator [talking about mixed numbers] and then, and why do we do that. That process of okay let's go back to the pies. Let's go back to the diagrams.

These comments seemed to indicate that there should be student mastery of understanding the "why" of the concept before moving on to the procedure. Roger (QCF) stated that the students have a tendency to move to the procedure. He shared that this benefits the strong student and can become somewhat of a limitation for students:

> But, you know for example, I am thinking about the division of fractions, I return to the fractions. The strong student passes automatically to the procedure even if they have not yet understood the representation. They are happy to have a procedure to use.

Further comments by Guillaume seemed to tie into the importance of students having a sound conceptual understanding and the capability of explaining why. For him, the concept of a fraction should transfer to other contexts as well as connecting to other fractions. He said, "When, I believe that someone knows what a half is, it can be a cup, it can be this, it can be that, since I can transform it, and divide it in two, at some point, in five, in six" (Guillaume, NBF). This idea of transfer connects to another idea that the New Brunswick teachers shared: connecting the teaching to real life, though all three groups discussed this element. For Guillaume, this was something important for making connections. He contended that if links were not made between mathematics and life, then the students would not remember what they were taught. In addition, it was important to "make as many links as possible with things we know" (Guillaume, NBF). He described the necessity for the mathematics to relate with the important domain of real life. He said, "Once we explained it, it was mastered [understanding of fractions], we offer the choice, 'choose what is most comfortable to you, and go.' We are more like in real life" (Guillaume, NBF).

Finding real-life opportunities for fractions was discussed by the teachers. Chloé shared how she used to cook in the classroom to allow the students the opportunity to measure using measuring cups (Figure 5.5). She shared how she was disappointed that she could no longer do this activity for policy reasons. Baking was something that Carine also talked about; however, she questioned how real such activities were for students. She commented, "Well, for recipes

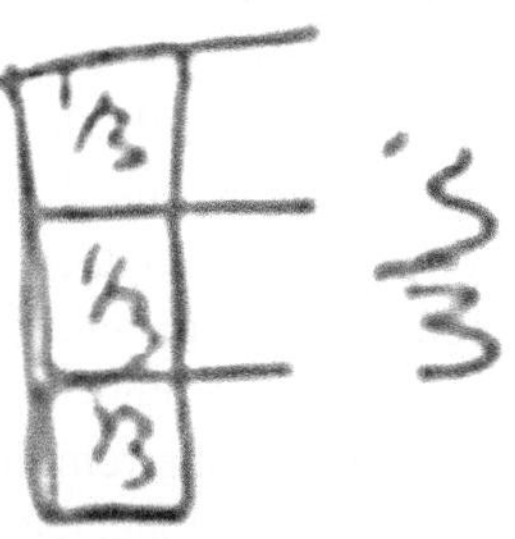

FIGURE 5.5. Representation of Fractions with Drawing of Measuring Cups.

it is not something that I do, but I think, there are lots of students for which it is not even something concrete using measuring cups" (Carine, QCF). Despite this, Irène countered this criticism and said, "I liked the idea of the cups. The imperial system, it is important in life." Consistent with the idea of using real measuring cups, Johanne contributed that she thought it was important for the students to have the opportunity to work with manipulatives. For her, this was especially the case for students who were transitioning from elementary to secondary school. She explained that there are interesting online resources or visual manipulatives. However, she cautioned that some of the students need manipulatives or need to work it out on a board.

Linking Fractions to Everyday Life

Tying into the idea of cooking, baking, and real life was a strong connection to food. Throughout the teacher discussions, there were numerous references to pizza, pies, and cakes. It was possible to see that these were easy examples for the teachers to conjure up even in their discussions with each other. Roger stated, "But on the board, if we have just, I don't know, a pie, and then after that we divide it in four, in five, because we are increasing the denominator, in six, in seven" (Roger, QCF). This is very similar to Johanne's description of how she ended a certain lesson talking about numerators and denominators. She recalled for the other teachers that, at this point of the lesson, she was talking about cake. For Guillaume, the link to food was especially an opportunity to help students make a connection with something they know. He shared that, in class, when he shows representations that are parts of a pizza, he goes the extra step to tell the kids that he would eat a piece in an effort to tie it to something real for his students.

The exchanges between the groups of teachers offered many opportunities to talk about food. Some teachers seemed somewhat cognizant of this fact. This can be seen by Guillaume, who said, "... but she returned at the end by talking about pizzas, about meals" (Guillaume, NBF). Camilla also made a similar comment about another teacher's practice when she made a comment that discussed how she was also showing pies to accompany a visual representation.

Another part of the discussion about context was about how a Quebec francophone teacher connected her lesson to music. This was appreciated by Danielle, who said, "I adored the comparisons, the links that she made with music" (Danielle, NBF). The teacher who implemented the lesson was teaching to a special group in a music concentration. The teacher used the musical notion of crescendo to have the students talk about increasing and decreasing order of fractions. Danielle applauded this when she added, "And, the crescendo it was super fun, because you have students in music. You capture them right away with something that is concrete for them" (Danielle, NBF). This remark again demonstrated the fact that teachers appreciated when lessons connected to something meaningful for students.

Coming back to the idea of interpretations of fractions, throughout the discussions the Quebec francophone teachers were the only teachers who discussed the ratio interpretation of fractions. This interpretation is where the teachers' understandings of fractions as ratios also begin to tie into pedagogical content knowledge. This can be seen as they talked about how an understanding of a ratio can help students determine what a fraction of something can be. Roger talked about how the idea of taxes on a total purchase helps students determine fractions such as three fourths. Roger (QCF) said:

> With the taxes, that is how we will present it. I have 15% each time that I spend one hundred dollars, I give 15 dollars. I spend 100 dollars, I spend 15 dollars. It really is to present it like that, three fourths. Each time I have four tokens, there are three aside.

This was something that Carine also used in her classroom when students were having difficulty determining three fourths of 48. She re-enacted a conversation with her students when they explained to her that they were okay doing three fourths of one by splitting the whole into four, and coloring in three pieces. To this she added, "Super, splitting it into four is what operation then?" (Carine, QCF). She explained that the students were comfortable talking about how they were dividing in four and taking three pieces, multiplying by three. She continued, explaining that although the students were capable of doing that, if you tried to ask them to find three fourths of 12, it was no longer instinctive to divide by four and multiply by three. She went on to explain how, with her students, they solved three fourths of 48 by drawing a rectangle to represent the 48 and then dividing it by four and coloring three of those pieces. With this, she said the students were fine determining that each piece of the rectangle was worth 12, but she emphasized that it was challenging and took time for the students to integrate the idea and be independent.

One last aspect of the mathematics discussed by the groups of teachers related to observations made about the numerator or the denominator. As already highlighted in the reflections about circles and rectangles, the use of rectangles to show the relationship between the denominator and the size of the pieces was appreciated. In the New Brunswick group, the teachers highlighted an activity done by a Quebec francophone teacher, where the goal was to organize fractions in order from smallest to largest. In this activity, Anne-Marie appreciated the choice of numbers for the fraction. She mentioned how this was beneficial to understanding what it means when either the numerator or the denominator is the same for two fractions. Guillaume agreed, stating, "She classed them [the fractions used in the exercise] by placing them such that they had to understand, the students, the sense of the largest numerator, the largest denominator" (Guillaume, NBF).

Curriculum Knowledge

In addition to real-life connections as discussed under the category of subject content knowledge, Chloé touched upon the importance that she attributed to making connections to prior knowledge. She shared that it was important to reactivate what students had previously learned. She considered this important when asking her students questions. In those moments, she explained that she always saw the same students raising their hands. In those situations, she provided hints, such as "do you remember when we…?" (Chloé, NBF). She explained that this would "wake them up a bit," (Chloé, NBF) and it would connect to prior knowledge. Finally, in contrast to talking about the beginning of a lesson, one teacher praised and recognized the way another teacher concluded a lesson. For Kayla from the Quebec anglophone teacher group, the culmination of the lesson incorporated many elements and made it concrete. She shared, "I liked the closing and I thought that it was the most concrete: teaches vocabulary, found the words properly, and visually showed it" (Kayla, QCA). In considering these last two comments, not only was it important to tie the lesson to prior learning, it was also important for it to recapitulate and end concretely.

Along with how the groups of teachers voiced the goals and approach that should be part of lessons, they also shared some limitations that connected to knowledge of curriculum. In this theme, two elements were not surprising: technology and time constraints. Although technology was considered an asset for some of the participating teachers, on occasion it also created obstacles. This was voiced by Irène from the Quebec francophone teacher group. She discussed how she used to use a technological tool to separate polygons, which now no longer works. This limited her options when representing fractions. She added that her drawing tool made it difficult to separate rectangles. Therefore, dividing a rectangle into equal parts was complicated when it was divided in more than three or four parts. Given how often the teachers appreciated the use of a rectangle over a circle in representing fractions, this limitation is unfortunate, and restricts the use of fractions like one seventh and complicates representations in class.

With regard to time constraints, the teachers felt that more class time was needed to allow for exploration of ideas. Guillaume elaborated that often the focus ends up just being on getting the answer. This is something that he expressed was difficult when navigating the program. He stated that often the emphasis is on results, but what he truly wished for was that teachers would have time to let students go and explore.

CONCLUSION

In closing this chapter, it was interesting to see how the participating teachers, by providing a window into each other's classrooms through videos, sparked discussions about how they thought they should teach fractions, highlighting some productive teaching practices. It was only by taking part in such conversations that teachers could engage in important "aha!" moments. An example of this was Kayla's (QCA) admission about falling into "the trap" of always showing fractions as circles and "always doing the pizza."

In exploring the exchanges of the three groups of teachers as they viewed video-recorded lessons on fractions, we can conclude that the results gave us a better idea of what is important for them in relation to teaching and learning fractions. From our analysis, we classified the ideas that the teachers contributed into three main categories using Shulman's (1986) framework: subject matter content knowledge, pedagogical content knowledge, and curriculum knowledge. We summarize our findings in the next paragraphs.

As for subject matter content knowledge, in conversations about the meanings of fractions, the teachers explicitly discussed three of the five interpretations of fractions: part of a set, as a ratio, and as a measure. The most common interpretation for representing fractions, part of a whole, was not explicit in the teacher exchanges. The teachers said that the part of a set interpretation of fractions can create a barrier when talking about the addition of fractions. As for the number line, it was considered by the Quebec francophone teacher group to be largely overlooked (Figure 5.3). For this reason, this group of teachers recommended a lesson that incorporated the number line as a choice of video to share with other teachers. In addition to discussing the importance of the numerator and the denominator, teachers communicated some big ideas about fractions. These ideas included the importance of the whole and the idea of equipartitioning. They also discussed a notion that they considered abstract for students: the number of parts necessary to recreate the whole.

As for pedagogical content knowledge, all three groups of teachers discussed the important role that visual representation has in teaching and learning fractions. The participating teachers asserted that visual representations should be explored before students get too involved in learning the steps, and before application. The group of teachers from New Brunswick contributed the idea that visual representations were a good way to assess student understanding and were necessary before the use of short cuts. An important aspect of visual representation was the idea of using a rectangle instead of a circle. Some teachers admitted focusing too much on the convenience of circles, and therefore, they did not incorporate enough rectangles. However, the rectangle was recognized as superior in certain situations, such as demonstrating the size of the pieces as the denominator increased.

The groups of teachers also touched on ideas of equivalency, decimals, and percentages. They viewed percentages and decimals as other ways of expressing fractions that were important for students to understand. For the group of teachers in New Brunswick, the decimal representation seemed to be more accessible to students. Operations on fractions were equally discussed. For this, the participating teachers focused uniquely on the operation of division, excluding the idea of adding two collections, discussed above, as being a challenge for students.

Finally, language was considered to be an important element of students' and teachers' capacities to communicate. This last discussion of language was an idea that was difficult to fit neatly in the mathematics of a fraction category, as it spilled over into goals and approach.

As for the discussions of goals and approach, there was a focus on understanding both the "how" and the "why" of developing conceptual understanding. As for the context, there was discussion of how teaching should incorporate as many links as possible, especially connecting to real-life. For younger students transitioning from elementary school, the use of manipulatives was recommended. In terms of manipulatives, there was also some discussion about using real-life applications like cooking and measuring cups, but some thought that this no longer was concrete for students. Music was another application that was appreciated from another teacher's lesson, although the participants thought that this was not appropriate for all students.

Finally, as for curriculum knowledge, two other elements that were considered important by the teachers were the connection to student's prior knowledge and the conclusion of a lesson. In this category, teachers touched upon two limitations that they said prevented them from achieving these goals. These were technology and time. In the case of technology, it sometimes limited the teacher's capacity to represent fractions in various forms. Not surprisingly, time was said to be limited given the expected results required.

As we mentioned previously, although studies on teaching and learning fractions appear to have been widely conducted, how teachers perceive teaching fractions is less prevalent. Although results of this study brought interesting aspects of teaching fractions into focus, we argue that this is only a starting point for further research. More studies on how teachers perceive teaching fractions should be conducted in various contexts to see how the results relate.

REFERENCES

Ball, D. L. (1993). Halves, pieces, and twoths: Constructing and using representational contexts in teaching fractions. In T. P. Carpenter, E. Fennema, & T. A. Romberg (Eds.), *Rational numbers: An integration of research* (pp. 157–195). Mahwah, NJ: Lawrence Erlbaum.

Behr, M. J., Harel, G., Post, T. R., & Lesh, R. (1992). Rational number, ratio, and proportion. In D. A. Grouws (Ed.), *Handbook of research on mathematics teaching and learning: A project of the National Council of Teachers of Mathematics* (pp. 296–333). New York, NY: Macmillan.

Behr, M., Lesh, R., Post, T. R., & Silver, E. A. (1983). Rational number concepts. In R. Lesh & M. Landau (Eds.), *Acquisition of mathematics concepts and processes* (pp. 91–126). New York, NY: Academic Press.

Chapin, S. H., & Johnson, A. (2006). *Math matters: Understanding the math you teach.* Sausalito, CA: Math Solutions.

Confrey, J., Maloney, A., Nguyen, K., Mojica, G., & Myers, M. (2009). Equipartitioning/splitting as a foundation of rational number reasoning using learning trajectories. In

M. Tzekaki, M. Kaldrimidou, & H. Sakonidis (Eds.), *Proceedings of the 33rd Conference of the International Group for the Psychology of Mathematics Education* (pp. 792–799). Thessaloniki, Greece: Aristotle University of Thessaloniki.

Da Ponte, J. P., & Chapman, O. (2006). Mathematics teachers' knowledge and practices. In A. Gutierrez & P. Boero (Eds.), *Handbook of research on the psychology of mathematics education: Past, present and future* (pp. 461–494). Rotterdam, the Netherlands: Sense.

Elbaz, F. (2018). *Teacher thinking: A study of practical knowledge*. London, UK: Routledge. https://doi.org/10.4324/9780429454615 (Original work published 1983.)

Kieren, T. E. (1993). Rational and fractional numbers: From quotient fields to recursive understanding. In T. P. Carpenter, E. Fennema, & T. A. Romberg (Eds.), *Rational numbers: An integration of research* (pp. 49–84). Mahwah, NJ: Lawrence Erlbaum.

Lamon, S. J. (2012). *Teaching fractions and ratios for understanding: Essential content knowledge and instructional strategies for teachers*. New York, NY: Routledge.

Rosar, D., Van Nieuwenhoven, C., & Jonnaert, P. (2001). Les fractions, comment mieux comprendre les difficultés rencontrées par les élèves? [Fractions, how to better understand difficulties encountered by students]. *Instantanées mathématiques*, *37*(2), 4–16.

Schwartz, J. L. (1988). Intensive quantity and referent transforming arithmetic operations. In J. Hiebert & M. Behr (Eds.), *Research agenda in mathematics education: Number concepts and operations in the middle grades* (pp. 41–52). Hillsdale, NJ: Lawrence Erlbaum Associates.

Shulman, L. S. (1986). Those who understand: Knowledge growth in teaching. *Educational Researcher*, *15*(2), 4–14.

Vergnaud, G. (1983). Multiplicative structures. In R. Lesh & M. Landau (Eds.), *Acquisition of mathematics concepts and processes* (pp. 127–174). New York, NY: Academic Press.

Vergnaud, G. (1988). Multiplicative structures. In J. Hiebert & M. Behr (Eds.), *Research agenda in mathematics education: Number concepts and operations in the middle grades* (pp. 141–161). Hillsdale, NJ: Lawrence Erlbaum Associates.

PART II

INTRODUCTION TO PART II

Seeking Similarities in Pedagogy

In Part II, we share outsider analysis of the data as we attempt to use teachers' utterances to understand how they conceptualize pedagogy. Each of us, though part of the research team, was not part of a data collection team, and thus we did not work directly with the participating teachers. Instead, we each examined transcripts produced from the fieldwork done by other members of the team. The dataset in each case consists of the transcripts of the auto-ethnographies (conversations as teachers selected videos of the teachers' own lessons to share with others) of the three anglophone focus groups: in Alberta, Ontario, and Quebec. Pragmatically, these transcripts were used by the authors because they were available to us in a shared first language (English) of researchers and participants and were among the first to be ready for analysis. In the next three chapters, we bring our outsider perspectives to the data shared by our insider colleagues.

One of the recurring challenges of much classroom research is the problem of how to say anything about teachers' or students' understandings of what they do, when these understandings are largely inaccessible to direct researcher observation. This problem arises at individual and social levels. At an individual level, it is not possible to have direct access to what research participants are thinking or feeling. Any research about thinking or feeling or other mental states must be mediated by language (or some other semiotic system). Common research meth-

Researching Pedagogy and Practice with Canadian Mathematics Teachers, pages 117–123.

ods include interviews, observations, and analysis of classroom interaction, all of which rely heavily on interpretation of what participants mean by what they say. However, participants' words carry all manner of cultural and linguistic baggage that comes with using a shared language system (Bakhtin, 1981). Moreover, a researcher's interpretation is shaped by the prior cultural and linguistic assumptions of the researcher. At a social level, it is not possible to directly interpret a culture from observing participants' behaviour. Much of what might be considered cultural is implicit, and members of a culture may not be aware of cultural patterns or assumptions (Garfinkel, 1967). And again, a researcher's interpretation of behaviors in terms of describing a culture or some aspect of culture is derived from the interaction between a researcher's own history and the situation he or she is observing. It is not possible to separate the researcher from his or her observations. It is, therefore, not possible to directly describe a culture, in the same way that it is not possible to directly describe individual mental states.

We have described this problem in methodological terms: how can we conduct research on pedagogical cultures if we cannot directly see them? In fact, teachers face this same problem every day in their work. It is not possible to directly access what students are thinking or what they know, and it is not possible to directly read students' cultures of mathematics, or music, or family simply by observing them.

The problem is not a new one, and various theoretical and methodological approaches have been developed in response, such as in the enactivist thinking that formed the basis for the Observing Teachers project. Looking specifically at methodological assumptions, Reid and Mgombelo (2015), for example, state: "Instead of asking questions about an external world, enactivism asks questions about observers and the worlds they observe" (p. 179). This position is related to the enactivist point that "the process of description of data is an interrelationship, in which researchers find themselves learning new things within a context which is partially of their own creation" (Reid & Mgombelo, 2015, p. 179).

This position has led many researchers to adopt a collaborative research stance, so that participants are explicitly involved in *bringing forth* the findings of the research. Davis and Renert (2013), for example, posit that teachers' understandings of mathematics are nuanced and emergent, and that participation in collaborative studies with colleagues can have a profound impact on the further development of their evolving pedagogy. They suggest that these knowledges "are tacit and can only emerge through participation in collective explorations" (p. 264).

The goal of the Observing Teachers project faces exactly the kinds of challenges we have discussed. How can we, as researchers, describe cultures of mathematics teaching that are in some sense true to teachers' perspectives, rather than reproducing our own assumptions about pedagogy. Whether we, the researchers, are closely familiar with classrooms in a particular region, or have never visited these classrooms, our reading of what is going on is derived from our assumptions and prior experiences of mathematics classrooms, teaching, and so on. For example, over the course of the project, the research team has spent much time discussing and debating possible definitions of mathematics pedagogy. Such discussions are important. Nevertheless, our observations about pedagogy are necessarily from an outsider position, at least to the extent that we do not work with the children, schools, or contexts in which the teachers who are participating in this project are teaching. How, then, can we say anything about the teachers' perspectives on pedagogy?

Recalling that "everything is said by an observer" (Maturana, 1987), the process of data analysis is an interrelationship, in which researchers find themselves learning new things within a context that is partially of their own creation. The changes that can be triggered in us, that is, what we can learn about the research context, are determined by our theories, beliefs, and biases. What we learn is determined by what we know (Reid, 1996). It is, therefore, helpful and important to provide some information about who we, the observers, are in the work presented in Part II and identify for the reader the lenses we bring to the data (although we acknowledge that you, the reader, are an observer too, and will bring your own lenses as you make sense of this writing).

AUTHORS' BACKGROUNDS

Richard Barwell is originally from the UK, where he completed an undergraduate degree in mathematics, a post-graduate certificate of education (mathematics), and graduate degrees in mathematics education. He taught secondary school mathematics in the UK, and taught mathematics, science, English, and various other things in a community school in northern Pakistan. He has worked at the University of Ottawa (in Ontario) since 2006 and has conducted several research projects in elementary schools in both Ontario and Quebec, including a multi-year ethnography of second language mathematics classrooms. Much of Richard's research uses discourse theories and discourse analysis combined with ethnography as a research strategy.

More specifically, Richard has written about the challenges of analyzing mathematics classroom interaction when researchers do not have direct access to students' or teachers' internal mental states, such as what they know, mean, or feel. In response, Richard has drawn on ideas from ethnomethodology, conversation analysis, and discursive psychology (see Edwards, 1997; Edwards & Potter, 1992) to examine how participants themselves deal with this problem of making sense of each other's words. An approach based on discursive psychology focuses on how the idea of *mathematical knowledge* or *mathematical knowing* is produced and deployed in mathematics classroom interaction (Barwell, 2013). From this perspective, the idea of knowing mathematics, as instantiated in various utterances about what participants know, is shaped and adjusted according to the particular situation.

Lisa Lunney Borden has never lived outside of Atlantic Canada. She completed all her formal K–12 schooling in New Brunswick and her university education in Nova Scotia and New Brunswick. She taught briefly in New Brunswick but spent most of her teaching career in Nova Scotia, having taught grades 7–12 mathematics at a Mi'kmaw First Nation community school for 10 years before moving to academia. Lisa teaches at St. Francis Xavier University in Nova Scotia, where she earned her undergraduate and graduate degrees in education. She completed her doctorate at the University of New Brunswick. She currently holds the John Jerome Paul Chair for Equity in Mathematics Education, with a strong research focus on supporting mathematics achievement for Mi'kmaw and African Nova Scotian students.

This project was a bit of a departure from Lisa's usual work, which often takes a critical focus attending to systemic inequities and ways to address social justice issues. Her work typically draws upon theoretical frameworks rooted in Indigenous knowledge systems that centre Indigenous ways of knowing, being, and doing that stem from her experiences working alongside Mi'kmaw community members for nearly 30 years (Lunney Borden, 2017). Situating her work in Indigenous research methodologies (Kovach, 2009; Smith, 1999), she often works collaboratively with teachers, students, and community members taking a decolonizing approach to mathematics education that supports not only student learning, but also the development of cultural identity. She is keenly interested in how mathematics emerges from community contexts, particularly when cultural knowledges serve as the starting point for learning (Lunney Borden & Wiseman, 2016; Wagner & Lunney Borden, 2015). She also focuses her attention on how the verb-based structure of Indigenous languages can transform pedagogical practices in mathematics and support greater success of all learners (Lunney Borden, 2011, 2013).

Ellen Carter has been educated solely in rural Nova Scotian schools and has worked as a mathematics teacher in the same province for about 10 years, primarily in middle school. Completing her undergraduate (mathematics) and two graduate degrees in education at St. Francis Xavier University, she now works as a part-time instructor and research assistant for Lisa while enrolled in the Ph.D. program at the same institution. Like Lisa, her work focuses on social justice in education, with a particular interest in youth who are in the care of the child welfare system. The critical and complexivist lens through which she approaches her own work, combined with her experiences as a teacher and teacher educator, contribute to her perspectives on teacher practice, mathematical pedagogy, and relational, structural, and systemic forms of power.

David Reid's teacher education and school teaching experience occurred in Quebec. He completed his doctoral studies, which included school-based research, in Alberta, and he was a long-time teacher educator in Atlantic Canada. This suggests that he has, at least partly, an insider perspective on teaching in three of the regions in this project. However, David attended school in the United States, not in Canada, and his teaching experience in Quebec is now two decades old. Therefore, in some ways, he is an outsider looking at the teaching that the teachers observed. His recent experiences working with colleagues on school-based research in England, France, and Germany have also affected how he sees mathematics teaching.

David's research has primarily focused on proof and reasoning, with careful attention to students' thinking and motivations. He has applied the idea of an "emotional orientation" (Maturana, 1988) to understanding how a need to prove arises in learners. This interest led to exploring the same idea in the context of teachers' motivations (see, for example, Brown & Reid, 2006). In parallel, David began to consider how classroom environments position learners in different ways, from a sociological perspective. These experiences mean he brings an eclectic mixture of theoretical frameworks, though predominate among them are ideas from Maturana and Varela (e.g., Maturana & Varela, 1980).

MULTIPLE OBSERVERS, MULTIPLE PERSPECTIVES

We believe that the three distinct approaches adopted in the chapters of Part II show how we can bring different lenses to transcripts and generate different ideas, highlighting how what is said is said by an observer. Indeed, a characteristic of an enactivist methodology is seeking out "multiple consensual contradictory perspectives" (Reid, 1996).

> Multiple perspectives can refer to many aspects of enactivist research. The most obvious is the participation of a number of researchers, each with her or his own agenda, theories, and background. … At the same time, we work with a common collection of data, about which we each reach conclusions related to our own interests and theories. These conclusions need not be parts of a single consistent whole. In fact, particular interpretations might be quite different. While some interpretations are not accepted by every member of the group, all interpretations must be explicable. That is, it must be possible to explain the conceptual structure in which the interpretation holds, even to others who may not see the world through that conceptual structure. I use the phrase "multiple consensual contradictory perspectives," where "consensus" is used to mean the explanation of interpretations in ways which make sense to others, to capture the important features of perspectives in enactivist research. (p. 207)

The three analyses in Part II are an attempt to arrive at such perspectives. We would expect some commonalities, as they are needed to make our interpretations sensible to others, but also some contradictions. The perspectives provide insight both into the perspectives of the observers, and into the data itself. Contradictory perspectives let us see what distinct stories the data can tell us, rather than trying to limit the story to a single narrative. In Part II, then, we present three distinct analyses of the auto-ethnographies of anglophone Ontario, Quebec, and Alberta teachers. To conclude this part of the book, we discuss the multiple consensual contradictory perspectives that emerge from this work.

In reading these chapters, we propose some questions that the reader may wish to consider:

- What resonates with you, and what do you see in your own practice?
- Are there key ideas that you think are missing?
- Which method of analysis do you think makes the most sense to understand middle school mathematics pedagogy in Canada?
- How might these three analyses taken together develop a clearer picture of middle school mathematics teaching in the Canadian anglophone context?

REFERENCES

Bakhtin, M. M. (Ed.). (1981). *The dialogic imagination: Four essays*. (C. Emerson & M. Holquist, Trans.). Austin, TX: University of Texas Press.

Barwell, R. (2013). Discursive psychology as an alternative perspective on mathematics teacher knowledge. *ZDM-International Journal on Mathematics Education*, *45*(4), 595–606.

Brown, L., & Reid, D. A. (2006). Embodied cognition: Somatic markers, purposes and emotional orientations. *Educational Studies in Mathematics*, *63*(2), 179–192.

Davis, B., & Renert, M. (2013). *The math teachers know: Profound understanding of emergent mathematics*. New York, NY: Routledge.

Edwards, D. (1997). *Discourse and cognition*. London, UK: Sage.

Edwards, D., & Potter, J. (1992). *Discursive psychology*. London, UK: Sage.

Garfinkel, H. (1967). *Studies in ethnomethodology*. Englewood Cliffs, NJ: Prentice Hall.

Kovach, M. (2009). *Indigenous methodologies: Characteristics, conversations, and contexts*. Toronto, Canada: University of Toronto Press.

Lunney Borden, L. (2011). The "Verbification" of mathematics: Using the grammatical structures of Mi'kmaq to support student learning. *For the Learning of Mathematics*, *31*(3), 8–13.

Lunney Borden, L. (2013). What's the word for…? Is there a word for…? How understanding Mi'kmaw language can help support Mi'kmaw learners in mathematics. *Mathematics Education Research Journal*, *25*(1), 5–22.

Lunney Borden, L. (2017). A journey to transforming mathematics education for Mi'kmaw students. In M. Battiste (Ed.), *Visioning a Mi'kmaw humanities: Indigenizing the academy* (pp. 253–286). Sydney, Canada: Cape Breton University Press.

Lunney Borden, L., & Wiseman, D. (2016). Considerations from places where Indigenous and Western ways of knowing, being, and doing circulate together: STEM as artifact of teaching and learning. *Canadian Journal of Science, Mathematics and Technology Education*, *16*(2), 140–152.

Maturana, H. (1987). Everything is said by an observer. In W. I. Thompson (Ed.), *Gaia: A way of knowing: Political implications of the new biology* (pp. 65–82). Great Barrington, MA: Lindisfarne Press.

Maturana, H. R. (1988). Reality: The search for objectivity or the quest for a compelling argument. *The Irish Journal of Psychology*, *9*(1), 25–82.

Maturana, H. R., & Varela, F. J. (1980). *Autopoiesis and cognition: The realization of the living*. Dordrecht, the Netherlands: Reidel.

Reid, D. A. (1996). Enactivism as a methodology. In L. Puig & A Gutiérrez (Eds.), *Proceedings of the Twentieth Annual Conference of the International Group for the Psychology of Mathematics Education,* Vol. 4. (pp. 203–210). Valencia, Spain: PME.

Reid, D. A., & Mgombelo, J. (2015). Survey of key concepts in enactivist theory and methodology. *ZDM-International Journal on ZDM Mathematics Education*, *47*(2), 171–183.

Smith, L. T. (1999). *Decolonizing methodologies: Research and indigenous peoples*. London, UK: Zed Books.

Wagner, D., & Lunney Borden, L. (2015). Common sense and necessity in (ethno)mathematics. In K. S. Sullenger & R. S. Turner (Eds.), *New ground: Pushing the boundaries of studying informal learning in science, mathematics, and technology* (pp. 113–128). Rotterdam, the Netherlands: Sense Publishers.

CHAPTER 6

"I DO THAT TOO"

Teachers' Constructions of Shared Mathematics Pedagogy

Richard Barwell and Lisa Lunney Borden

This chapter reports our analysis of focus group discussions involving three groups of anglophone mathematics teachers, each from a different region of Canada. The teachers discussed videos of their own teaching to identify examples of mathematics teaching that were typical of their region. The focus of our discourse analysis was on how each group of teachers constructed a regional pedagogy through their interactions. We highlight several features of pedagogy for each group. Comparing the three groups' constructions of mathematics pedagogy, we found more similarities than differences.

The overall goal of this project was to identify regional pedagogies. In this chapter, we seek to understand how teachers themselves identify pedagogical values they share. To that end, we take a discursive approach to examining aspects of pedagogy observed by teachers. The title of the chapter, "I do that too," comes from one of the transcripts of a discussion involving the Ontario teachers observing each other's videos. Richard was struck by this comment as indicative of what Ontario mathematics pedagogy might be for the participants. When participants say things like "I do that too," they are showing recognition of a practice they have in common with other participants. As we shared emerging ideas during research

Researching Pedagogy and Practice with Canadian Mathematics Teachers, pages 125–144.

team meetings, Lisa noted that she had seen similar utterances in transcripts from other regions, and thus, we agreed to explore this emerging idea in more depth.

In this chapter, we look at what aspects of pedagogy are recognized by teachers within each anglophone region as being in common. We focus our analysis on teachers' constructions of common pedagogy in data from the first stage of the project gathered in the anglophone regions in which mathematics teachers in each region observed and discussed videos of each other's teaching. We believe that our approach of attending to how teachers discursively develop their own sense of shared pedagogy can help us understand how they are thinking about their own pedagogy, at least in the context of observations of videos of each other's teaching in a research setting.

USING A DISCURSIVE APPROACH TO UNDERSTAND PEDAGOGY

Our general orientation to discourse analysis is informed by discursive psychology. From this perspective, our focus is not on what teachers know (in their heads), but rather on how the idea of *mathematical knowledge* or *mathematical knowing* is produced and deployed in mathematics classroom interaction (Barwell, 2013). The idea of knowing mathematics, for example, is instantiated in various utterances about what participants know, and is shaped and adjusted according to the particular situation.

The enactivist position with respect to observers, outlined in the introduction to this section of the book, is more or less identical to the discursive psychology position that, instead of asking questions about the mental states of teachers or students (which exist in a world external to the researcher), asks questions about participants or members of a culture and the worlds that they describe. Research using this kind of approach has, for example, shown how students may be positioned as *not knowing* through particular interactional sequences (Reis & Barwell, 2013), or how students' mathematical explanations are constructed as relevant (Ingram, Andrews, & Pitt, 2018). In discursive psychology, discourse is pervasive, so that all research is reflexive, including research texts (Ashmore, 1989; Barwell, 2009). In both discursive psychology and enactivist research, there is an awareness that the researcher is not simply recording the objective world, but rather is producing, or bringing forth, a world or text. This position is related to the enactivist point that "the process of description of data is an interrelationship, in which researchers find themselves learning new things within a context which is partially of their own creation" (Reid & Mgombelo, 2015, p. 179).

Recall that in the first part of the project, middle-school mathematics teachers in each region prepared three 20-minute summary videos representing three lessons for each participant, resulting in 12 videos per region. The four participants in each region met together to review each other's videos, with the goal, set by the researcher, of choosing three videos to send to teachers in the other participating regions. The idea was that the three chosen videos would in some way represent the mathematics pedagogy of the region (even if such a goal were somewhat simplistic). The teachers' discussion as they formulated their choice of three videos was recorded; this was the data we analyzed for the results we present in this chapter.

Through their observations of the videos of each other's teaching, the participating teachers themselves clearly identified aspects of their mathematics teaching they shared. Through analysis of the teachers' discussions during the selection process, the teachers' discourses of mathematics pedagogy can be examined in their own terms. Of course, these discourses must be understood within the context in which they arose: as part of a research project, and, perhaps more importantly, with some concern for what teachers from other regions might find interesting. Within that context, it was possible to examine how teachers negotiated among themselves the commonalities in their practice and so constructed or brought forth a sense of a common pedagogy, as well as differences. Because we employed a discursive perspective for our analysis, we needed to adopt a modified definition of pedagogy consistent with our theoretical assumptions. This is because our goal was not to identify elements of regional pedagogy directly, but rather to identify what participants saw as common pedagogy for themselves within their region. For this chapter, our discursive definition of mathematics pedagogy was simply *teachers' ways of talking about teaching*.

Having noticed that participating teachers did indeed observe commonalities in the teaching within their region, as initially indicated by comments like "I do that too," we designed a systematic strategy for analysis. To ensure a manageable amount of data and a systematic sampling of the data, we selected the transcripts of the anglophone teachers' first meeting in each region, which typically included their discussion of four videos out of the 12 in their region (the rest were discussed in subsequent sessions). The three regions featuring anglophone teachers were Ontario, Alberta, and Quebec. The sessions were facilitated by a different researcher in each region with some differences in approach. (See Chapter 1 for a detailed account of the data collection methods.)

We began our analysis by identifying instances in which teachers explicitly observed other teachers' practice as being like their own. For example, as detailed later in this chapter, when Ontario teacher Jordan shares how much student talk happens in the classroom, Ellen replies, "Yeah, I teach with my door closed all the time because we are loud." To which Jordan responds, "Yeah, yeah, and so are my classes, yeah. I couldn't have it any other way." The two teachers indicated that they agree this is something they both value. We sought out these moments of agreement, which could be identified in various ways. Most obviously, participants explicitly expressed agreement with words like "yeah." Other indications of shared practice included more elaborate formulations of agreement (as in "I do that too"), or affirmation through reformulation, or adding to or developing what another participant said. For example, in one exchange, one teacher says, "you're writing a test" and her colleague responds, "you're worried." The response affirms the first comment by extending it.

For such instances, we extracted the surrounding sequence from the transcript in order to have the full discussion relating to the identified aspect of common practice. We then selected two representative sequences for each region, to see what they revealed about how teachers talk about their teaching. In particular, we focused on two aspects of teachers' talk: how students and students' activity is talked about, and how teachers and teachers' practice is talked about. Each utterance was examined to identify whether it referred most prominently to students' activity or to teachers' activity. We focused on these two aspects because they occurred frequently in the participants' discussions. Participants devoted considerable time to talking about what students did or should do, often in direct relation to what they did or should do as teachers. In some cases, utterances were divided, with some parts referring to students, and others to teachers. The transcript was then organized in a chart, with one column for parts of utterances referring to students and another column for parts of utterances referring to teachers. From the chart, we looked at each column to develop a global sense of how the participants talked about the activity of students and teachers. This approach allowed us to get a sense of participants' own ways of jointly producing a sense of regional pedagogy.

As we worked through our analysis, the participants' discussion in each region often brought to mind aspects of regional curriculum and policy. In some cases, for example, participants used particular language that seemed to be derived from emphases apparent in regional curriculum documents. These connections were sometimes explicit, or else were made by us, the researchers, during our analysis. For this reason, it is helpful to contextualize our findings by summarizing some relevant features of the mathematics curricula in the three regions we examined.

Regional Curriculum Discourses

The discourse of mathematics pedagogy in North American research is dominated by constructivist reform approaches, as instantiated in the National Council of Teachers of Mathematics (NCTM) *Standards*, the US Common Core, and by provincial curricula in Canada. The assumptions of this broad approach are well known, and emphasize the role of the learner in making sense of mathematics, the importance of working with what students can already do, the utility of suitable tools, and the value of discussion and reflection in exploring mathematical ideas (NCTM, 2000). This approach has become highly politicized, with sometimes ill-informed debate in national news media polarizing between "discovery learning," a shorthand characterization of the reform approach, and "the basics," taken to refer to direct instruction of methods, procedures and facts (see, for example, Abtahi & Barwell, 2019). These debates tend to focus on "results"—i.e., student performance in standardized tests or national performance in surveys such as the Programme for International Student Assessment (PISA). As a result, mathematics pedagogy is reduced to a simplistic choice between two broad approaches, neither of which is a realistic representation of what is actually expected in schools.

In Canada, provincial bodies write curriculum outcomes that set out what students will learn at each grade level. Although there are often provincially approved texts, teachers have autonomy to make decisions about lesson materials they will use, how they will help children meet outcomes, and how they will assess students. Many provincial and district level education authorities provide opportunities for ongoing professional learning for teachers to focus on particular areas they have identified as priorities. For example, in recent years, a mathematics strategy in Ontario supported teacher learning in mathematics through collaborative inquiry teams and professional development relating to matters of classroom discourse and structuring effective classroom pedagogy. Such initiatives are quite common in most provinces, though the topic for learning may vary considerably based on local priorities (e.g., Bruce, Esmonde, Ross, Dookie, & Beatty, 2010).

The Ontario mathematics curriculum for Grades 1–8 is based on problem-solving. The introduction to the curriculum document presents a summary of Polya's problem-solving cycle (Polya, 1945) and the rest of the document includes examples of problems that could be used for the different topics at each grade. The curriculum also details a set of mathematical processes that run through the various content requirements, including problem-solving and communication, with process expectations stipulated for each grade. The curriculum summarizes several points regarding effective approaches to teaching mathematics (Ontario Ministry of Education, 2005, pp. 24–25). Among other things, teachers are encouraged to:

- Use an investigative approach based on rich problems;
- Use well-chosen, meaningful, relevant contexts;
- Draw on students' prior knowledge, experiences, and interests;
- Focus on key principles;
- Use a range of tools;
- Listen carefully to students' reasoning to inform instructional decisions; and
- Develop students' communication skills in mathematics.

Alberta curriculum documents are similar to those of Ontario in that they emphasize mathematical processes that are important for mathematics teaching and learning. Teachers are reminded in the documents that children should construct their own meaning by engaging in a variety of mathematical experiences, including problem solving, using manipulatives, and engaging in meaningful discussions (Alberta Education, 2007, pp. 1–10). Teachers are encouraged to:

- Integrate processes into each content strand;
- Use a balance of understanding, recalling, and applying math concepts;

- Increase students' fluency through problem solving, reasoning, and connections;
- Help students develop concepts through concrete, pictorial and symbolic representations; and
- Acknowledge diverse learning styles and abilities.

The Quebec curriculum has a much greater focus on processes than Ontario and Alberta, and emphasizes process expectations more than content expectations. Students are primarily expected to develop mathematical competencies in the processes of communicating, reasoning, and problem solving using the content as a means to do so. Furthermore, content expectations span grade levels from introduction, through development, to consolidation, as opposed to having a specific content expectation for each grade, as do the documents for Ontario and Alberta. Teachers are expected to create an atmosphere where students can take an active role in their own learning through problem solving and explorations that allow them to reason mathematically, justify their thinking, and communicate what they have learned (e.g., Government of Quebec, 2006).

TEACHERS' REGIONAL PEDAGOGICAL DISCOURSES

Our analysis revealed how, for our participants, mathematics pedagogy was about creating the conditions for mathematical thinking. More specifically, the teachers typically focused on three aspects of their teaching and students' participation:

- Mathematical processes; such as conjecturing, justifying, and asking questions;
- Mathematical knowledge; such as calculation procedures, mathematical vocabulary, and arithmetic facts; and
- Affective concerns; such as managing students' struggle, supporting weak students, and dealing with students' anxieties.

In each region, there were different emphases and different orientations to these three aspects, with some points being common across regions. It is important to see these foci as *emic*, meaning internal to the culture of the participants. We are reporting aspects of pedagogy that were part of the participants' discussions of their teaching, rather than our own outsider observations.

Ontario

The first extract from the Ontario teachers, shown in Figure 6.1, illustrates how classroom talk was a shared feature of mathematics pedagogy for the participants. As mentioned earlier, we particularly focused on moments of explicit recognition of common features, such as when participants said things like "I did that too." The moments we identified in this way are italicized in the following examples. In line with our analysis, we present the transcript of each extract with columns

	Teachers	**Students**
Ellen:	I wanted and I think so I was able to hear all of that and then know where to go from there in terms of their understanding. And the question that I gave them at the end was just to see if	them to be doing a lot of talking in that particular lesson they did a lot of discussion of what they knew they knew how to operate then, with those numbers, because the [Grade] 7's wouldn't have done that yet. The [Grade] 8's obviously should have that background knowledge, so...
Researcher:	Okay, anything resonate with your teaching or particular things you do, or any comments you want to make?	
Jordan:	Considering student voice, there's a lot of opportunity for and you know I think about our classrooms at our school we *see quite a bit as well in my own work,* generally.	the kids to do some thinking and some talking, and contributing We do a lot of talking. In fact, probably we spend about 1/3 of our time talking at least, about stuff and testing out ideas, and conjectures if you will.
Ellen:	*Yeah,* I teach with my door closed all the time because	we are loud
Jordan:	*Yeah, yeah, and so are my classes, yeah.* I couldn't have it any other way	If the class is too quiet it feels unhealthy
Ellen:	*It does, yeah,* or you're writing a test	
Trista:	*You're worried*	
Jordan:	*Yeah. So yeah, I think that's what I see in common with the work that we do*	it's the talk. It's the discussion and kids proposing something and justifying and others perhaps countering and so....

(*continues*)

FIGURE 6.1. Ontario Anglophone Teachers' First Meeting to Select Videos, Extract 1

	Teachers	Students
Ellen:		*Yeah, yeah,* and having that opportunity for debate. You know it's okay not to agree
Jordan:		*Yeah*
Ellen:	And that's why I like those kinds of questions because they allow for that opportunity to see things in different ways	

FIGURE 6.1. Continued

showing attention to teachers and teachers' activities (on the left), as distinguished from students and students' activity (on the right).

This sequence shows how three of the teachers recognized each other's accounts of the importance of talk and student activity in their mathematics classes. Students are described as making healthy noise, and as engaging, through talk, in thinking, contributing, testing out ideas, conjecturing, proposing, justifying, countering, and as having a voice and opportunities for debate. Teachers are described as desiring this level of talk, and as facilitating the students' thinking through listening, "know[ing] where to go," and asking questions that "allow for that opportunity to see things in different ways." The behaviors described by the participants broadly relate to mathematical processes; the participants described various mathematical behaviors, and associated teacher facilitation of such behaviors, which they saw as desirable and as shared in their practice. For example, in response to Ellen, Jordan said "there's a lot of opportunity for the kids to do some thinking and some talking, and contributing and you know I think about our classrooms at our school we see quite a bit as well in my own work." Thus, for Jordan, a shared feature of their two classrooms was students engaging in thinking, talking, and contributing. The subsequent turns reinforced this identified similarity and highlighted that, for both participants, their classrooms are, as a result, "loud" and that they see this as a good thing. Trista affirmed Jordan's comment that a quiet classroom "feels unhealthy," saying "you're worried." This comment was very brief, however, and can be understood as arising from the general orientation to agreement that underlies any conversation. Saying "you're worried" achieves group cohesion by displaying an understanding of the exchange between Ellen and Jordan. Hence, all three teachers jointly produced a vision of a classroom in which students are active participants. In the three teachers' account, this activity is observable and apparent from the amount of noise that is generated.

In the extract from the Ontario data, shown in Figure 6.2, participating teachers discussed the importance of allowing students to struggle with challenging mathematical ideas.

	Teachers	Students
Trista:	I like	the struggle
	and *I do that too*—letting them	struggle
	letting them	get frustrated when the student made that sound, that "arghhhh"
	I love that. Those are my favorite sounds. Those and the	"wahh"
Ellen:	Yes	
Trista:		The "I get it sounds"
	But, yeah, I feel *that's* a very important thing to let them do.	
Steve:		*Sometimes it's the "arrr."*.. *then it's the "ahhh"*
Trista:	*Yeah*	
Steve:	And then you hope	they *don't* forget that
Trista:	*Yeah*, because *we're* so quick to help. And *that's* been a part of my own struggle as a teacher because you're so wanting to just get in there and help them and save them but the stepping back and giving them time	
	and realizing that often	they will get it and get more out of it when
	you let them	do that
	let them	struggle
Ellen:	Yeah	
Jordan:	I think it's variable though, I mean	some kids, depending upon their level of self-esteem and how they're feeling,
	they do need to be saved every now and then	because they're in the process of going under
	so you have to kind of like haul them back up to the top and give them a little bit of air and then you know, get them	to try swimming again
	So it's a real, it's a delicate balancing act between intervening and deciding to back off and it's highly variable, it really depends upon the individual	
Trista:	And that speaks to knowing your students and if you only teach them math, some teachers do, taking the time	
Ellen:	I'm experiencing that this year for the first time, actually	

FIGURE 6.2. Ontario Anglophone Teachers' First Meeting to Select Videos, Extract 2

A second common feature of their teaching identified by the Ontario participants related to the way they deal with students who are struggling. The four teachers described struggle as sometimes valuable, although they also related it to students' anxiety (not shown in the extract) and needing to be "saved." Struggle, as discussed by the teachers, is an affective issue (anxiety is about how students feel about the struggle, rather than the struggle itself). The participants, therefore, discursively constructed or brought forth a world in which students and teachers are understood in terms of how they feel about mathematics. The participants' accounts included some vivid affective description created through voicing students' expressions of struggle, frustration, and success: they make "aggh" sounds and "I get it" sounds.

The key moment in this exchange was Trista's summary: "And that's been a part of my own struggle as a teacher because you're so wanting to just get in there and help them and save them but the stepping back and giving them time." She related a struggle for her as a teacher with the struggles of students with mathematics, what we might call a *meta-struggle*. The teacher's meta-struggle was about how to manage students' struggles. She placed in tension a teacher's desire to help—to "save" students—with students' need for space ("stepping back") and time to work on the mathematics.

Jordan offered a different interpretation, but nevertheless accepted the nature of the meta-struggle. In Jordan's variant, students do sometimes need to be "saved," the choice of word being developed into a broad metaphor of drowning. This intervention began by challenging some aspects of Trista's account, but concluded by offering a very similar formulation of the meta-struggle: "So it's a real, it's a delicate balancing act between intervening and deciding to back off and it's highly variable, it really depends upon the individual." This formulation allowed Trista to agree, aligning herself with Jordan by providing the development that for teachers to make decisions in the classroom, they need to know their students.

In this sequence, then, the participants produced a second commonality in their teaching of mathematics, through a process of mutual recognition: an attention to supporting a degree of struggle in students' work while not allowing students to struggle too much. This affective account of mathematics pedagogy can be linked to the Ontario mathematics curriculum's emphasis on an investigative or problem-solving approach; in problem solving, methods and solutions are not always immediately apparent, and hence, a tolerance of uncertainty and struggle is important.

Alberta

The theme of students' struggles with mathematics also arose in Alberta. The extract in Figure 6.3 shows how the teacher participants used similar language to the Ontario teachers to talk about this aspect of students' mathematical behaviour, although their account of their own practices is less similar.

As with to the Ontario teachers, the Alberta teachers wanted their students to become comfortable with the struggle that is needed to learn mathematics. Again, this was an affective concern ("just comfort"), although the teachers related it to

	Teachers	Students
Andrea:	I like	my kids to take risks, and be willing to take risks, and struggle, and that's, you know, you do a lot of struggling
Brenda:		And they don't like struggling
Andrea:	No, but you, you know, you can train them	*and they do get comfortable with it, but it's okay to say in front of everyone something that might not be correct,* and we'll go from there. That idea. I guess it's, um, just comfort
Maryanne:		*Yeah, and it's not even in front of the classroom, it's just when they're working on work, how do I do this, how do I do this, not*—what do you think you do, you know, I really
Andrea:		*Try something*
Maryanne:		*Yeah*
Andrea:		*Try something different*
Maryanne:	Not giving them the answers, but letting	*[them] think through it*
Researcher:	*Yeah, so, I mean that's, yeah, let them*	think through it
Maryanne:	And *just pointing them in the right direction* instead of telling them	
Researcher:	So, would you say, Maryanne, that you're a person that kind of points them	
Maryanne:	Yes. *Just ask the questions until I kind of get the response I'm looking for, to guide them*	
Brenda:	*Always answer a question with another question*	
Researcher:	So, that's part of the way that you teach	
Maryanne:		Absolutely, because a lot of kids just want, they want you to tell them how to do it
Brenda:		They just want the right answer

FIGURE 6.3. Alberta Anglophone Teachers' First Meeting to Select Videos, Extract 1

problem-solving behaviors, such as having students think about how they know something or by posing a question. Andrea framed this exchange in terms of students taking risks in mathematics, and the other participants developed this theme together. The teachers and the researcher then produced an account of how teachers deal with this struggle: "not giving them the answers," letting students "think through" the problem, "pointing them in the right direction," and guiding students with questions. Through this part of the discussion, the participants built a coherent joint account of mathematics pedagogy. Maryanne and Brenda then provided a justification for this pedagogy: "A lot of kids just want … you to tell them how to do it. They just want the right answer." This sequence began with a framing position that students should be willing to take risks, but which they do not like. The teachers then produced an account of a pedagogy designed to promote risk-taking, which is justified in terms of students wanting to be told "how to do it." The last point can be heard as a description of non-risk-taking behaviour. The teachers' pedagogy is designed to get students comfortable with struggle, by "pointing them" in the right direction through asking questions.

There are similarities with the Ontario teachers' discussion about when to "save" students from struggling too much, but there is no sense of the meta-struggle that they invoked. We noticed some slight differences between the two groups. While in Ontario, in some circumstances teachers adopt a strategy of "stepping back and giving them time," in Alberta the participating teachers spoke of "*pointing them in the right direction.*"

In a second passage from the Alberta teachers' discussion, another aspect of mathematics pedagogy emerges, prompted by the editing of a video in which much of the seatwork done by students was cut. In the Alberta part of the study, full-length video recordings of mathematics lessons were edited into 20-minute reductions by a research assistant according to some general guidelines and without consultation with the teachers, which was not the case in other regions. On seeing one such reduction (see Figure 6.4), the teachers noticed the general absence of footage of the students working and discussed why they believed that the deleted footage showed an important part of what happens in their classrooms.

In this extract, we see that a period of time when students are working on problems is a part of what Alberta teachers identified as important mathematics pedagogy (it is not clear whether this work is individual or in small groups). We see how the teachers value the time when students are working on problems in their classrooms. The three teachers seemed to be in agreement that their role during this time is an important one. Their account of their practice during problem solving was in terms of students having "a chance to do it" and "input" into solutions. This role for students emerged over several turns of talk in the interview. This role clearly mattered to the teachers because they point out its absence from the videos several times. In particular, it mattered to the teachers that students contribute. Andrea summarized this point: "It's about what the students are bringing, so that idea that they're contributing, which I think, yeah, we don't see a lot

	Teachers	**Students**
Brenda:	Well, I think that notion of chunking the activities where we're doing a little bit of problem solving, and	then getting the kids—giving the kids a chance to do it
Brenda:	and then having that walk-around time	
Andrea:	I was going to say, but you don't see (*laughter*)	
Researcher:	It really is there!	
Brenda:	It is there	
Researcher:	So, sort of the problem	
Brenda:	Yeah, the problem	
Researcher:	Giving the kids, the kids time	
Brenda:	And then the activity, or whatever it was, because you had a very specific activity the kids were doing, and that got chopped out, right? Um, I'm thinking *even just my little checks for understanding* where they're, um, you know, like still completing. And then Andrea's, um, you know per—*your personal solutions where the kids were*	
Andrea:		*Yeah, I was gonna say, student input*
Brenda:		*Yeah, student input for solutions*
Researcher:	And some of this we can alre—probably easily identify from video.	
Andrea:	*Mhm*	
Brenda:	*Mhm*	
Researcher:	So what would you like to see from—or what would you like to exemplify	
Andrea:	Uh, well, I think that that's kind of it for me, is the whole idea that, um, it's about	what the students are bringing, so that idea that they're contributing, which I think, yeah, we don't see a lot in the videos, because that's more their work time.

(*continues*)

FIGURE 6.4. Alberta Anglophone Teachers' First Meeting to Select Videos, Extract 2

	Teachers	Students
Researcher:	*Mhm*	
Andrea:	But I think it's a little bit in all of them, at least.	
Brenda:	*And it's more about the learning, not about the teaching, I mean*	
Researcher:	Oh, okay	
Maryanne:	Yeah, I don't really—I'm thinking mine, just, I'm giving them a problem and I'm having them work on it, checking for understanding and then coming back and kind of going through it. So, correcting misconceptions on, probably, a wider scale. Because if they were more individual, I would help those students as I was walking around, but when it's kind of like seeing the common misconceptions, coming back together as a group and doing more of the teaching then, I guess	

FIGURE 6.4. Continued

in the videos." Maryanne described her role in these parts of her classes in terms of collecting and correcting misconceptions, either individually or with the whole class. Her description of this role seemed to clarify what "teaching" means to her, because it followed Brenda's comment that "it's more about the learning, not about the teaching." This exchange thus clarified what the teachers described as "teaching:" setting problems, collecting students' ideas (their "input"), and clarifying their misconceptions. For Maryanne, teaching was about her active guiding and clarifying misconceptions with students, something she does more with the group: "Coming back together as a group and doing more of the teaching then." The word *misconceptions* implies a concern with mathematics content, which nevertheless remains rather implicit.

Quebec

As in Ontario and Alberta, the Quebec teachers displayed a concern with students' affective responses to learning mathematics. The extracts presented from Ontario and Alberta focused on the need to be comfortable with struggling in mathematics in the context of problem solving, expressed in terms that often point to affective responses. In Quebec, in their discussion of the videos of their teaching, rather than struggle, the teachers talked about the importance of students being comfortable with making mistakes as seen in Figure 6.5.

The three teachers in this exchange produced a common sense of an aspect of mathematics pedagogy that involves being a model of mathematical behaviour for students. The kinds of behaviors they see themselves modelling included making mistakes, learning from mistakes, and celebrating mistakes. Tamara affirmed this

	Teachers	Students
Camilla:	When we make a mistake, it allows the students and it, how we react to our mistakes … is so important. I think we need to teach students that it's okay to make mistakes. That we can laugh at our mistakes, that we can celebrate our mistakes. That we learn from our mistakes and I think how we handle our own personal mistakes, teaches them	how to handle theirs whether it's a typo or a calculation error or whatever in the process where the kids can pick up on it and, and it's okay, so that I think that when kids see, oh wow
	she can make mistakes too well that's okay then I can make mistakes	
Tamara:	*But, also to move on from making mistakes,* right?	
Camilla:	*Oh, of course*	
Tamara:	*Yeah, it's okay to make mistakes*, but then what do you do after?	
Kayla:		It's important for them to know too that like really school is a very small like microcosm of the world where there's [not] one right answer. Life doesn't happen like that. There is very rarely one right answer
	so to really appreciate the diversity of	what they come up with, like how they approach it
	that's what I think elevates an exceptional lesson. To, to allow yourself to deviate from the plan. So to model that you're doing the same. And, and to take risks with the kids because that's what I think we want them to do. I think it's about relinquishing control a little bit	

FIGURE 6.5. Quebec Anglophone Teachers' First Meeting to Select Videos, Extract 1

view, but suggested that accepting mistakes in mathematics needs to be followed with something more. This provided an opening for Kayla, who recast Camilla's account in terms of a broader set of mathematical values and teacher behaviors. An important mathematical value is "there is very rarely one right answer," while teacher behaviors include deviating from the plan, taking risks, appreciating students' different ideas, and relinquishing control. Thus, modelling the acceptance of mistakes as part of doing mathematics is generalized as a case of a general pedagogical approach that involves taking risks with students. This approach was described in terms of students having positive agency, contributing different ideas that teachers should appreciate.

The formulation of this account of mathematics pedagogy was produced fairly unproblematically; the participants did not challenge each other's characterizations or offer strongly different accounts. Camilla's account was confirmed by Tamara and then extended by Kayla. The participants seemed to recognize each other's contributions as fitting into a shared sense of their teaching.

In a second passage from Quebec (see Figure 6.6), prompted by one of their own videos, the teachers discussed specific ways to help students learn vocabulary. Learning and teaching mathematical vocabulary were frequently discussed by the Quebec teachers.

In this extract, Camilla and Kayla, along with the researcher, produced an account of a specific aspect of their teaching. Both teachers recognized that vocabulary is important, and gave an account of how it can be treated in their teaching. In this account, they drew attention to vocabulary. They asked the students if they

	Teachers	Students
Researcher:	So, earlier, you said an example when you introduce a new concept and you have your student talk and introduce a new vocabulary. Do you think it's possible to use 'why' in every lesson, every activity in class?	
Camilla:	Yes, because even when you're introducing new vocabulary, when you're talking about rotations, and enlargements, and reductions, and congruence. Well, why congruent? Even though it's new vocabulary to that chapter, it's not new math vocabulary to them, or even enlargement, why enlargement? The word 'large' is in enlargement so that, you know, so, even if it's a totally new word	
		usually their prior knowledge from another subject, in that case, it being English language, for them was an automatic answer to them. Okay, enlargement, large. Reduction, reduce. You know?
	So, I think you can always ask the whys	

FIGURE 6.6. Quebec Anglophone Teachers' First Meeting to Select Videos, Extract 2

	Teachers	Students
Kayla:	*I think, the same thing* the other day. We were introducing monomials and mono, that's a scary word, and I always, as much as possible, I always try and teach the meaning of new words, I break it into the parts and components of the words, whether we're looking at the prefixes or suffixes of the word, we talk about the individual parts. *So I said, so look, we wrote the whole definition then I said "okay, now let's go back, let's look at the word if we forgot, is there a clue in the word that's going to help you remember again?"*	
	So, and I was amazed	how many hands went up to say "well, mono means one." It's great. This is the first time I've ever had a group that knew that right away
	So, I said, "how do you know that?"	"Oh, in our history class, monotheistic means that they only believe in one god"
Researcher:	*Mmmm*	
Kayla:	I went, "nice, okay"	so there's a connection there that they knew that like the back of their hands, so now for monomial, we're talking mono, one, monomial is a number or term. So then, they put that together and ... there we go
	So, then we were able to move forward and that was a connection, but it would have been so easy to just, monomial, here's the definition, that's it, that's all. And not ask the next question	
Camilla:		And if her kids see on the final exam, "express your answer as a monomial," her students are going to go "one, mono, mono, one, one term." Whereas the students who didn't hear that will go "I can't do this question, I have no idea what I have to do." Right? Right away they're in because of the 'why,' why the monomial? It's not a random word, right? that serves no purpose

FIGURE 6.6. Continued

recognized key terms, and prompted students to notice components of words in order to make connections and be able to make sense of potentially unfamiliar words in the future. The teachers' account constructed students as being unlikely to know some vocabulary and unlikely to be familiar with some prefixes or other components—Kayla expressed surprise, for example, that her students were familiar with the prefix *mono*.

The importance of vocabulary is linked, in particular, to evaluation. Camilla referred to the "final exam," thus framing the teaching of mathematical vocabulary in terms of its value for assessment purposes. Indeed, this remark suggests that, for these teachers, the "final exam" is a significant aspect of their pedagogical thinking. We see that these teachers valued the importance of making connections to what students already know when learning mathematics vocabulary. Their account also described the role of the teacher as one of asking questions and prompting potential connections. Students were expected to engage in the discussion and make meaningful connections of their own.

DISCUSSION AND CONCLUSIONS

Our analysis for this chapter focused on aspects of pedagogy that were recognized as being in common by teachers within each anglophone region. Our analysis revealed several such aspects, which we argue are suggestive of features of regional mathematics pedagogy as recognized by teachers in their discussions about videos of the teaching within their region. During their discussions, teachers often displayed mutual recognition of common features, from which we can infer that these aspects of teaching were recognizable to the participants. We acknowledge that our samples are small, but we see emergent ideas that the teachers in each region recognize as shared views of pedagogy. Having adopted a discursive perspective that privileges participants' own accounts, we have some confidence that these features are meaningful for these teachers, without in fact needing to know exactly what they look like for ourselves.

What features of mathematics pedagogy did teachers identify? In all three regions, a concern with students' encountering and overcoming difficulties in the process of learning mathematics was apparent, with teachers in Ontario and Alberta highlighting the importance of students learning to struggle during problem solving and accepting such struggle as part of doing mathematics. For the teachers in Ontario, this concern was related to a tension about when to step in to help students who were struggling, and when to leave them to work through it. In Alberta, there was a sense that teachers would guide students through such struggles, such as by asking questions. In Quebec, the teachers talked about the value of making mistakes and the need for them to model making mistakes as a normal part of doing mathematics. They related this aspect of their teaching to a broader account of giving students some control and acknowledging students' contributions.

The role of language emerged in different ways in each region. In Ontario, the teachers emphasized the importance of student talk in learning mathematics, and as a result, celebrated the noise levels in their classrooms ("We are loud"). They were clear that such talk has mathematical value, particularly in developing reasoning through "proposing something and justifying and others perhaps countering." This led another teacher to frame this point in terms of debate and emphasize that "it's okay not to agree." This is a different emphasis from the Quebec teachers' discussion, not only because the Quebec teachers focused more on vocabulary, but also because they talked about mistakes. There is perhaps a subtle difference in perspective here, in which Quebec teachers paid more attention (in their discussion) to whether students' ideas were correct or not, and see students' ideas as "input." By contrast, in Ontario and Alberta, the teachers wanted students to try things out and not be focused on "how to do it." In Quebec, it is the teachers who take risks; in Ontario and Alberta, it is the students. This difference suggests that in Quebec, teachers saw their role differently than teachers in Ontario and Alberta.

The overall accounts of mathematics pedagogy in each of the three regions are related to the curriculum expectations in each region, although there is no clear correspondence that explains any differences. All three groups stressed the importance of problem-solving and of mathematical processes in general, and the extracts we have selected speak to these aspects of the curricula. Curiously, in the province in which processes are most strongly emphasized—Quebec—the teachers seemed to have the pedagogy that was least process-oriented. The teachers in Ontario and Alberta discussed students' engagement with mathematics at length and seemed keen on helping students have a positive relationship with the subject. In Quebec, the teachers displayed an approach that was more focused on right (and wrong) answers and the learning of mathematical vocabulary. The differences were not great, however, being more a question of emphasis than of strongly divergent ideas about teaching mathematics.

These findings contribute to our understanding of anglophone regional pedagogies in Canada. They highlight, for each region, some common practices and perspectives. We do not claim to have developed a comprehensive account of each regional pedagogy, not least because only four teachers participated from each region. What we can be confident of, however, is that the commonalities that emerged from our analysis are meaningful and recognizable for participants as our analysis was entirely based on what participants commented on themselves. A further contribution is that our findings suggest that teachers in the three regions included in this analysis highlight commonalities in similar ways and for similar practices. It was striking, for example, that classroom interaction was a point of shared attention in all three regions. This observation suggests that, while the particularities of practice may have a regional flavor, teachers across the three regions have much in common. This claim, which must be regarded as tentative, is perhaps not surprising, and may imply that regional variations in mathematics pedagogy are a matter of difference in emphasis, rather than of significant differences in classroom practice.

Finally, the approach to analysis we have adopted for this chapter shows the contingent, situated nature of teachers' accounts of their teaching. Teachers build on each other's contributions, generally producing a joint account that all participants seem to recognize. There was rarely any strong disagreement, with participants displaying a strong orientation to agreement. This orientation suggests that teachers are adept at this kind of conversation, working together to build a shared perspective. As such, we believe that this approach could be a significant research tool for those attempting to understand teachers' pedagogical beliefs.

REFERENCES

Abtahi, A., & Barwell, R. (2019). Mathematical morality tales: Mathematics education in Canadian newspapers. *Canadian Journal of Science, Mathematics and Technology Education, 19*(1), 48–60.

Alberta Education. (2007). *Mathematics: Kindergarten to grade 9 program of studies.* Retrieved from: https://education.alberta.ca/media/3115252/2016_k_to_9_math_pos.pdf

Ashmore, M. (1989). *The reflexive thesis: Wrighting sociology of scientific knowledge.* Chicago, IL: University of Chicago Press.

Barwell, R. (2009). Researchers' descriptions and the construction of mathematical thinking. *Educational Studies in Mathematics, 72*(2), 255–269.

Barwell, R. (2013). Discursive psychology as an alternative perspective on mathematics teacher knowledge. *ZDM-International Journal on ZDM Mathematics Education, 45*(4), 595–606.

Bruce, C. D., Esmonde, I., Ross, J., Dookie, L., & Beatty, R. (2010). The effects of sustained classroom-embedded teacher professional learning on teacher efficacy and related student achievement. *Teaching and Teacher Education, 26*(8), 1598–1608.

Government of Quebec. (2006). *Quebec Education Program: Preschool education and elementary education.* Quebec, Canada: Government of Quebec.

Ingram, J., Andrews, N., & Pitt, A. (2018). Making student explanations relevant in whole class discussion. In J. N. Moschkovich, D. Wagner, A. Bose, J. Rodrigues Mendes, & M. Schütte (Eds.), *Language and communication in mathematics education: International perspectives* (pp. 51–63). Cham, Switzerland: Springer.

National Council of Teachers of Mathematics. (2000). *Principles and standards for school mathematics.* Reston, VA: NCTM.

Ontario Ministry of Education. (2005). *The Ontario curriculum grades 1–8: Mathematics.* Retrieved from http://www.edu.gov.on.ca/eng/curriculum/elementary/math18curr.pdf

Polya, G. (1945). *How to solve it.* Princeton, NJ: Princeton University Press.

Reid, D. A., & Mgombelo, J. (2015). Survey of key concepts in enactivist theory and methodology. *ZDM-International Journal on ZDM Mathematics Education, 47*(2), 171–183.

Reis, G., & Barwell, R. (2013). The interactional accomplishment of not knowing in elementary school science and mathematics: Implications for classroom performance assessment practices. *International Journal of Science and Mathematics Education, 11*(5), 1067–1085.

CHAPTER 7

SIMILARITIES AND DIFFERENCES IN PEDAGOGY

A Thematic Analysis

Lisa Lunney Borden and Ellen Carter

This chapter presents a thematic analysis of the transcripts of the anglophone regions of the study from an outsider perspective. In our efforts to seek out regional differences in pedagogy, we actually began to see many similarities across regions through our analysis and coding. We first describe the themes we determined were shared across regions and then explore some additional themes that seemed to be unique to the three regions. We conclude with some discussion of the limitations of our outsider analysis and some insights into future areas to be explored.

In this chapter, we discuss our approach to analyzing transcript data in search of regional pedagogies. We first present an overview of our thematic analysis procedure, followed by a description of themes that emerged through this process. Each theme is augmented with quotes from research participants, both to highlight the identified theme for readers as well as to share examples of passages selected and coded through the analysis process. All participant names are pseudonyms assigned to individuals. Although the overall goal of the study was to identify regional differences in pedagogy, as we explored the data we also noted some striking similarities across regions. We became keenly interested in what

Researching Pedagogy and Practice with Canadian Mathematics Teachers, pages 145–156.

was common in pedagogy across all contexts as well as what might be unique in a specific region. In this chapter, we share our findings.

We first analyzed the data using open coding (Merriam & Tisdell, 2016), an emergent, inductive, cyclical coding technique whereby the codes are derived from the text. We examined only the transcripts of anglophone teachers observing their own group's videos of their teaching in three regions: Alberta, Ontario, and Quebec. We were attempting to reveal pedagogy, that is the "'implicit cultural practices' of teachers... not taught explicitly in schools of education or written down in textbooks [but which] reflect an implicit cultural logic" (Tobin, Hsueh, & Karasawa, 2009, p. 19). Hence, we read each transcript individually and identified ideas or quotes that highlighted beliefs, tacit knowledge, customs, or habits that guide practice, flagging and labeling these elements in passages of the transcript that might be relevant to the study. We then shared and discussed our individual observations to create a shared picture of the data.

The open codes were then further analyzed through analytical coding (Richards, 2015), the process of grouping the developed open codes into an evolving structure based upon our ongoing interpretation of the data. Comparisons were made between transcripts and codes within each teacher group, which informed the development of thematic categories to capture recurring patterns into which data items could be sorted. Together, we discussed emerging themes and how we might categorize them. All quotes were then copied into a spreadsheet and organized into key themes for each region. A secondary spreadsheet was then created to align regions by column and key ideas by rows to assist with the detection of commonalities and differences among regions. We observed considerable similarities across the three contexts as well as some regional differences.

NATIONAL SIMILARITIES

Among the three regions, similarities in the data emerged that we classified as teacher as facilitator, classroom community, active learning, curriculum connections, and pedagogical practices. Quotes are provided to highlight these resulting themes.

Teacher as Facilitator

In all regions, teachers viewed themselves as facilitators of learning and often contrasted this with traditional direct instruction approaches. In Alberta, the discussion and beliefs were shared among two participants:

Maryanne: That's what I did, too, but, I think in all of these cases, we're trying to get them to give as much as they can.

Brenda: We're the vehicle, we're just sort of the—we're more the facilitators, right?

Maryanne: Yeah.

Teachers, here, developed their shared understanding to confirm each other's pedagogical beliefs and then they named this underlying view as being one of facilitation, a notion further examined in Chapter 6. The participants' view of the role of teacher as facilitator is consistent with social constructivist approaches in curriculum (Ernest, 1994).

Bruner (1986) describes the constructivist teacher as one who offers appropriate tasks and opportunities for dialogue and guides the focus of students' attention. As a result, teachers must be knowledgeable about the typical sequence in which students develop mathematical ideas, about the current individual structures of students in the class, and must possess skills to structure the intellectual and social climate of the classroom (Clements & Battista, 2002). The social constructivist view recognizes the importance of social interaction through communication amongst students and with the teacher (Ernest, 1994). Reflective of this view, Jordan, an Ontario teacher, supposed that "Teaching math, or any subject for that matter, extends beyond the content material itself and it's about your role as facilitator, affirmer. It's a very complex task. It's beyond just content." Cobb (1988) supports this idea in saying that there is much more required of a constructivist teacher than simply teaching by imposition. In Quebec, Camilla indicated her perceived value of the extra effort in terms of efficacy, saying, "I see us as facilitators. Teachers are facilitators, we go in and we give them the tools they need, we give them direction, we give them structure, and if you have the right environment, then they will learn on their own." Teachers facilitating student learning in this way are seen as providing opportunities for students to develop the tools and skills to be independent, lifelong learners. The specific attention given not only to the teaching strategies, but to the environment in which students are expected to learn, leads to the second national similarity: classroom community.

Classroom Community

Moving beyond teacher actions, we again see an emergent social constructivist philosophy at play in the way teachers highlight the importance of classroom community, which includes the environmental structures and practices that encourage students to develop and maintain a mathematical community. This echoes Ernest's (1994) notion that social constructivism includes more than the social interactions, rather it extends to the materials used, the discourse of the classroom, and the interactions of content and communication. Teachers in all regions talked about the importance of creating a safe space for students to make mistakes and learn from each other, emphasizing the need to share their thinking and make meaning together. For example, an Ontario teacher, Steve, when discussing his own lesson reported,

> Or I would tell them, you know, "go and talk to some other groups and see if you can get some other ideas.... You know if you're getting really frustrated or you can't

> come up with any ideas on your own then feel free to walk around and check out what other people are working on—talk to other people."

This idea of building classroom community by encouraging students to talk with each other was reiterated in Alberta, and extended to include the importance of developing student comfort in sharing solutions and strategies openly. One Alberta teacher, Brenda claims: "You know, you really have to build the trust and the safety of being public like that."

In all regions, teachers considered and explicitly discussed expectations surrounding the sharing of student strategies and solutions in class, and specifically around making mistakes. Teachers value students sharing their mathematical thinking and consider the consequences of classroom environment in ensuring safety for students to do just that. It also points to the importance placed on conceptual understanding and thinking processes over procedures and final solutions. For example, two teachers from Quebec commented:

> Tamara: We always talk about a safe classroom environment, right? And we mean physically safe, of course, but we also mean a place where we respect the classroom, respect each other, respect the property, the people, that part of what you're saying would be to create that safe environment where good, bad, good answers, bad answers, every answer is welcome, every answer is encouraged and, you know, create that safe place for that...
>
> Kayla: And I think when we put more emphasis on the how and the why, it doesn't become about the answer anymore. So, really, I don't actually care what your answer is, I care how, how did you get there, what were you going to do to get there, and what was your thinking.

Another moment of discussion in Quebec focused on the acceptance of mistakes. As stated by Camilla:

> I think we need to teach students that it's okay to make mistakes. That we can laugh at our mistakes, that we can celebrate our mistakes. That we learn from our mistakes and I think how we handle our own personal mistakes, teaches them how to handle theirs.

This premise was mirrored again in Ontario with one participant, Ellen stating:

> The importance of making mistakes and so we spent a lot of time at the beginning of the year just building that comfort zone, and kids get anxious about sharing because they don't want to be wrong, and we talked about how mistakes drive learning and seeing things in a certain way and then changing your thinking actually builds more brain, more neural pathways in your brain than it does if you're getting it right all the

> time. So, I hope that sinks in and I hope that develops. It's a little bit too early to see how it's going to help but I'm hoping that will help build that confidence.

Teachers from each of these regions discussed the importance of not only the environment, but also the teacher's role in helping students develop a healthy relationship with mistakes. This focus also built upon the notion of teacher as facilitator, in particular, a facilitator of active learning.

Active Learning

Many teachers discussed the idea that students need to take an active role in their own learning, grapple with ideas, overcome challenges, and take responsibility for their own understanding. Kayla, a member of the Quebec anglophone focus group, articulated her experiences and beliefs in initiating this response from students:

> When you start digging deeper, yes, for many of them, it's uncomfortable, it's difficult and frustrating sometimes. But, as you continue to move them … I think they not only gain more intellectually, but they gain more in confidence because they feel that they overcame an obstacle.

Maryanne, an Albertan teacher, expressed student responsibility as a top priority: "I think my biggest one is just being an active participant when they're learning new content." Finally, in Ontario, Steve built on the role of teacher as facilitator to include beliefs around student learning by saying:

> Well and again, I like to give them some hints if they're getting stuck but at the same time, like several of you said, I don't want to tell them because if I tell them, like I say to them, "If I tell you, then who's done the math? Who's doing the work and the thinking?"

This notion of active participation is, again, reflective of a social constructivist view of education. Teachers valued student experience in constructing their knowledge, which was similarly reflected in their views of the role of the teacher.

Curriculum Connections

Within the theme of curriculum connections, the belief in the importance of moving away from traditional, abstract, procedural understanding was raised once more. Sub-themes emerged of real-world contexts for learning mathematics, conceptual understanding, and connections across mathematical concepts. In Quebec, Kayla expressed: "To memorize something, a rule, it doesn't attach to anything in their brain if they don't know why. If they really have an understanding of why, they don't have to memorize the formula, they understand it." Building on this notion, in Ontario Jordan expressed, "Math has to have context, just like everything else. It's absolutely meaningless unless you know what it means."

In Alberta, the recorded fraction lesson prompted Andrea to share: "I try to find real life connections wherever I can, so, what are you doing with these fractions, like, what's the purpose of multiplying them?" Aligned with the National Council of Teachers of Mathematics' (NCTM, 2000) *Principles and Standards for School Mathematics*, this focus on meaningful learning outlines the principle of learning as a need for students to "learn mathematics with understanding, actively build new knowledge from experience and prior knowledge" (p. 20). It further proposes that effective teachers design experiences and lessons that respond to, and build on, prior knowledge.

Pedagogical Practices

We use the term *pedagogical practices* to encompass many of the teaching practices that participants expressed as ways they believe learning happens in the classroom. These included group work, guided instruction, questioning, limiting lecture, listening, and circulating in the classroom, and are connected to the teachers' views of their role as facilitators of learning. Ellen, an Ontario focus group participant, noticed that there was a shared practice among all teacher participants in the math class and stated:

> We've talked about how we start together and then they get down to group work, we're constantly circulating, nobody's sitting behind a desk at a computer. Those student conversations, with either individual students or groups, was all pervasive—we all did it.

Building on the principles of teacher as facilitator and active learning, the physical manifestation presents itself as a teacher circulating while students actively work together. In Alberta, Maryanne noted a similar commonality amongst the region's teachers: "We do try to spend as much [time] as we can, the kids actually doing something, versus us talking." Quebec participants also contrasted their practices with direct instruction; for example, Camilla stated: "There's more to instruction than being up there and showing and demonstrating, you know? Students need to explore on their own." More explicitly, in a publication of the NCTM (2000) it is suggested that teachers should use worthwhile tasks, but also must utilize a range of pedagogical practices, such as highlighting aspects of tasks, organizing and orchestrating the work of students, challenging students with a variety of levels of expertise, and supporting students without eliminating the challenge by taking over the process of thinking.

Within this same theme of teachers as facilitators of active learning, specific strategies were highlighted by teachers in all regions. In Ontario, Jordan spoke of the use of questioning: "It's a question of just posing a question and letting their response just kind of sit there and ripen a little bit, and then you pose the next question so you're not necessarily giving them the answers but you're posing the question, the answer comes from them, then you kind of build on that for your next step." Questioning was also underscored in Quebec, specifically the use of open-ended questions, which again draws upon conceptual versus procedural knowledge. Tamara argued in favor of open-ended questions stating, "I'm asking you open-ended questions to get different opinions and different views." Mary-anne, an Albertan teacher, touched upon questioning in a synthesis of important ideas across our identified themes when she said, "It's kind of that environment, but I think we're seeing what they know, we're looking for what they know, looking for misconceptions, but also I think we're asking them to try to make them a part of it. Engage them." This final quote reflects not only the creation of a safe environment and the use of questions, but also denotes a form of ongoing assessment to inform teaching practices.

REGIONAL DIFFERENCES

Despite many consistencies across the regions, we also noted themes that were more prominent in specific regions. We noticed a dominant theme in Alberta around student engagement; in Ontario, the focus was on differentiation and planning for student success; in Quebec, participants focused on the importance of math talk and developing the vocabulary of mathematics.

Alberta: Student Engagement

We observed frequent use of the word *engagement* throughout all Albertan anglophone focus group meetings. We differentiate student engagement as distinct from active learning, detailed above. Active learning focuses on the autonomy and actions of students; student engagement is centered on motivation. In this sense, for Albertan teachers, engagement is not so much about the task design as it is about connecting to students' interests so they feel motivated to continue with their learning. An example of the key theme of student engagement can be seen in the following quote from Andrea, who notes the degree to which the teachers focus on engaging students: "Yeah, I am really concerned with covering the whole curriculum, most of my energy probably goes into, how am I going to engage the kids? How am I going to make them interested enough to learn this well?" Student engagement was addressed throughout conversations, particularly around planning for students, which requires a teacher knowing students and having ways to incorporate their interests into lessons. Brenda shared, "I just use it [things that relate to students] for engagement purposes." Furthering this conversation, the teachers gave an example of speed texting as something that interested students and explained how using that information as the data for a graphing unit could spark greater engagement. Teachers made explicit that student engagement is necessary to ensure that mathematics curriculum is covered, and students are moved from learning specific methods and focusing on solutions to understanding mathematics conceptually.

The significance of student engagement in Alberta was similarly noted and is more deeply explored in Chapter 2. We wondered if engagement was a dominant idea in the Alberta curriculum document. However, a search of the document found only one use of the word in a sentence describing the importance and development of student engagement in problem solving (Alberta Education, 2007). Since that time, the term *engagement* has been increasingly permeating professional learning experiences for teachers in Alberta. In a more recent document produced by Alberta Education (2017) to guide the development of new programs of study, there is considerable emphasis on the importance of student engagement and planning for such engagement.

Ontario: Differentiation

In Ontario, teachers in the focus groups described a recognition of the individuality of students' needs. Although discussions of differentiation were not restricted to Ontario, it occurred more commonly in the conversations of this region. In response to observing a lesson, Ellen explained that in her practice:

> If it's half the class, I'm going to recap and probably do it with everybody because it's a good review for the other group, but I might put a different task on the screen for those who do get it so that they can move on. And then I'm going to reassess it to make sure that the kids that didn't get it the day before get it today.

A sense of teacher responsibility was communicated. It became apparent that not only did the teacher need to ensure that all students had tasks that were relevant and met their needs, but also, the teacher needed to follow up and assess outcomes so that further differentiation could take place, if required. Other teachers spoke of the importance of open-ended tasks, multiple entry points, student choice, and having students work in groups to capitalize on the strengths of individual students. All of these practices and perceptions were fundamentally connected to an awareness of students as people gathered in a common space, more than as members of a group. It is notable that "The Capacity Building Series," produced by The Literacy and Numeracy Secretariat to support leadership and instructional effectiveness in Ontario schools, provides a resource for teachers entitled "Differentiating Mathematics Instruction" (2008). Furthermore, having supported professional learning for the Ontario Literacy and Numeracy Secretariat, Lisa (co-author of this chapter) was aware that differentiation had been a focus of professional learning experiences for regional lead teams supporting mathematics education more broadly in the province.

Quebec: Vocabulary

In Quebec, teachers described mathematics as a language with a specific set of sophisticated terms. The belief in the importance of teachers to both teach and model effective mathematical communication was shared amongst the participants in this group. This focus on language built upon the shared themes of understanding their role of teacher as facilitator and their need to support students' conceptual understanding, which was identified previously. Kayla stated, "We're going to use the proper vocabulary, so yeah, we can brainstorm all these different ideas, but we're going to bring it back to how do we say that in our most sophisticated math language." Throughout a focus group meeting, Kayla noted: "All of our discussions have really centered around [how] we need to be vocabulary specific and if we're holding the kids to an expectation that they're going to communicate using those higher-level vocabulary terms, we need to model that. And if we're not going to model that, neither are they." Provincial documentation for mathematics outlines three core competencies in Quebec, the third of which is "Communicate using mathematical language" (Gouvernement du Québec, 2004). Key features of this competency are becoming familiar with mathematical vocabulary, making connections between mathematical and everyday language, and interpreting or producing mathematical messages. We see this influence echoed in the words of the teachers. It may be important to consider that Quebec is the only province included in this analysis with French as the official language. Language is political in Quebec, where laws are in place to protect the primacy of French language. Because the teachers were anglophone, the broader complexities associated with language in the province may play a role in making language considerations unique to this region.

DISCUSSION

Through a thematic analysis we were searching for "'implicit cultural practices' of teachers ... not taught explicitly in schools of education or written down in textbooks [but which] reflect an implicit cultural logic" (Tobin et al., 2009, p. 19). We focused on identifying patterns of meaning within transcripts from each region by examining what was discussed by the teacher participants. We identified themes through coding and interpreting the transcripts, and through the ways in which the teachers themselves discursively formed some shared meaning (exemplified in the provided quotes).

A major theme for each region was identified: student engagement in Alberta; differentiation in Ontario; and vocabulary in Quebec. Although none of these themes were exclusive to any one region, we believe that these themes are representative of what was expressed most strongly in each. We recognize that the teachers were focused on these topics at the time of the study, and thus, their comments provide only a snapshot into the complexity that is mathematics pedagogy. None of these concepts happen in isolation in a classroom, and no theme can provide a full picture of pedagogy. Many of the themes are connected and interdependent. For example, what teachers say about having students work in groups is dependent on establishing a safe learning community and is related to the design of the tasks that students are asked to do in collaborative groups. How these ideas, values, and practices are combined and enacted in complex environments will differ and will further influence future pedagogy and practices. It is "not just the sum of its parts, but the product of the parts and their interactions" (Davis & Simmt, 2003, p. 138).

Although we focused on what the teachers discussed through the transcripts, we also acknowledge the importance of what is absent. When we (both researchers and participating teachers) pay attention to one thing, naturally other things are not given the same attention. Similarly, given the implicit cultural logic denoted in our view of pedagogy, there are opportunities for some key ideas to be taken for granted, and therefore unspoken in this context. For example, noticing that there is an emphasis on differentiation in Ontario, it may be an indicator that inclusion is a focus in those classrooms. Inclusion is not explicitly discussed, but the practice of differentiation may be a way to address an underlying/unspoken value of inclusion.

Looking at the emerging themes that are shared by the teachers across the country, we see, potentially, what may be Canadian pedagogical values: teacher as facilitator, classroom community, active learning, curriculum connections, and pedagogical practices. Although we recognize that our own backgrounds and perspectives influence our interpretations of the meanings attributed to teacher conversations, through our analysis we see evidence of shared views on creating classroom communities, focusing on classroom discourse, valuing conceptual understanding, and wanting to create independent thinkers who learn through doing. Should we walk into any middle school mathematics classroom in Canada, we expect that these practices and beliefs may be things that we might notice.

We are left with a few questions worthy of further exploration. First, many teachers expressed a belief that they were unique, or part of a small group of teachers who were engaged in these sorts of practices, and contrasted themselves with other teachers (outside the focus group) whom they believed were still engaging in more traditional direct instruction approaches. For example, Andrea from Alberta stated the idea that there might be some common pedagogy was intriguing to her, "I think that I don't even teach that similarly to the other people in my school, so … that just kind of piqued my interest." We wonder how teachers' views of other teachers' lessons are shaped. Exploring the first-hand or second-hand methods through which teachers create their understanding of what is happening in other classes is worthy of further investigation.

Second, we also wonder about the regional differences that were identified, and how these might be influenced by, or are related to provincial initiatives around professional learning. We noted influence from provincial documents in some instances; however, we believe there could be a more thorough examination of how teachers take up provincial initiatives and enact them in their classroom practice.

Third, we question whether teachers' pedagogy was affected by participating in this research, particularly if a shared pedagogy was created among the groups that differed from earlier individual views. Perhaps in future studies, pre-focus group conversations with individual participants would be useful as a means to identify the role that the focus groups played in shaping teacher beliefs.

From the transcripts of conversations, it is difficult to discern the level of congruence between teacher pedagogy and practice. Further inquiry into the alignment between teachers' pedagogical beliefs and actual practices may help to explain the role that teachers play in student achievement, though we recognize that this is a complex and difficult issue to investigate.

Overall, the significance of this analysis is that it begins to form a picture of what a typical Canadian middle school math classroom might look like. It allows us to recognize some of the things that teachers are valuing, and to give language and labels to important pieces of pedagogy. That being said, we recognize that we only had access to what teachers said about the tapes they viewed and not the tapes themselves. Had we watched the tapes, we may have noticed different things. For example, we might have seen teacher moves that are so normal for teachers that they become taken for granted, and are not talked about by teachers. We acknowledge that our picture is incomplete, but we think that our findings give us some perspectives with which to examine other classrooms. For example, what does differentiation, vocabulary, or student engagement look like in different classrooms—in the provinces in which those themes were most applicable and elsewhere? Do these ideas resonate in other parts of the country, or are these teachers unique?

REFERENCES

Alberta Education. (2007). *Mathematics kindergarten to grade 9 program of studies*. Retrieved from: https://education.alberta.ca/media/3115252/2016_k_to_9_math_pos.pdf

Alberta Education. (2017). *The guiding framework for the design and development of kindergarten to grade 12 provincial curriculum.* Retrieved from: https://education.alberta.ca/media/3575996/curriculum-development-guiding-framework.pdf

Bruner, J. (1986). *Actual minds, possible worlds.* Cambridge, MA: Harvard University Press.

Clements, D. H., & Battista, M. T. (2002). Constructivist learning and teaching. In D. L. Chambers (Ed.), *Putting research into practice in the elementary grades: Readings from journals of the National Council of Teachers of Mathematics* (pp. 6–8). Reston, VA: National Council of Teachers of Mathematics.

Cobb, P. (1988). The tension between theories of learning and instruction in mathematics education. *Educational Psychologist, 23*, 87–103.

Davis, B., & Simmt, E. (2003). Understanding learning systems: Mathematics education and complexity science. *Journal for Research in Mathematics Education, 34*, 137–167.

Ernest, P. (1994). Social constructivism and the psychology of mathematics education. In P. Ernest (Ed.), *Constructing mathematical knowledge: Epistemology and mathematics education* (pp. 62–71). Abingdon, England: RoutledgeFalmer.

Gouvernement du Québec (2004). Québec education program: Secondary school education, Cycle One. Retrieved from: http://www.education.gouv.qc.ca/fileadmin/site_web/documents/education/jeunes/pfeq/PFEQ_presentation-premier-cycle-secondaire_EN.pdf

The Literacy and Numeracy Secretariat. (2008). *Differentiating mathematics instruction* (Secretariat Special Edition #7). Retrieved from: http://www.edu.gov.on.ca/eng/literacynumeracy/inspire/research/different_math.pdf

Merriam, S. B., & Tisdell, E. J. (2016). *Qualitative research: A guide to design and implementation*. San Francisco, CA: Jossey-Bass.

National Council of Teachers of Mathematics. (2000). *Principles and standards for school mathematics*. Reston, VA: Author.

Richards, L. (2015). *Handling qualitative data: A practical guide.* London, England: Sage.

Tobin, J., Hsueh, Y., & Karasawa, M. (2009). *Preschool in three cultures revisited: China, Japan, and the United States.* Chicago, IL: University of Chicago Press.

CHAPTER 8

SEEKING PATTERNS IN PEDAGOGY

A Topical Analysis

David A. Reid

A comparison of the topics discussed by anglophone teachers from Quebec, Ontario, and Alberta while watching videos of their own teaching reveals some differences that may warrant further study. Differences were also observed in the topics discussed directly after watching the videos and during discussions at other times. Two teaching related topics were discussed more often in Ontario and Alberta than in Quebec, and two topics related to learning and to the institution of schooling were discussed more often in Quebec than in the other two regions. Three different topics related to learning and to the institution of schooling were discussed more often in Ontario than in the other two regions. These results are only indicative but suggest directions for additional study.

The research question for the Observing Teachers project, "How do pedagogies in middle school mathematics in regions of Canada differ?," can be addressed in many ways. In Chapters 2–7 we have shown some of the range of possible approaches. When I considered the data analysis, I debated a number of approaches. It was clear that I wanted some way to identify differences and similarities in what the teachers said, a common strategy in enactivist research. This meant that I wanted a method that would not be too fine grained, as in the details there were

Researching Pedagogy and Practice with Canadian Mathematics Teachers, pages 157–186.

bound to be many differences and few similarities. At the same time, I did not want an approach that involved me abstracting very general categories from the data. I felt that would tilt the balance too far towards revealing my own interests and too far away from saying something based on the data. In the end, I settled on examining topics of discussion, focusing on what the teachers discussed, without trying to infer how they felt about the topics of their discussion. I felt that the simple fact that they had chosen a topic indicates some valuing of it.

METHOD

The data I analyze here are drawn from the auto-ethnographies of the anglophone teachers in Quebec, Alberta, and Ontario. In each case, I divided the transcripts into episodes based on breaks imposed by the researcher and coded for the topic of discussion. Details for each transcript, and an observation related to each can be found in the appendices of this chapter. Here, I present the analysis from each region separately before comparing them.

Teaching is complex, and so any discussion of teaching necessarily addresses some aspects of teaching and neglects others. The topics addressed reflect a teacher's pedagogy, even before a particular position is taken. For example, referring to how students are grouped indicates that the topic of grouping is significant, whether preference is expressed for pairs, small groups, whole class instruction, or some pattern of combining groupings.

Codes

Transcripts of the focus group discussions were analyzed sequentially by region: Quebec, Alberta, and then Ontario. As such, the number of codes used grew in relation to the conversations within each region. The topics used in coding the transcripts are listed in Figure 8.1. The key words in parentheses were used as indicators and reminders of the nature of the topic. Codes added during Alberta analysis are denoted with *italics*, and those included through the Ontario analysis are *underlined*. After some reflection, I decided not to revisit the transcripts coded earlier using the new codes. I felt that a reanalysis would introduce only minor changes to the coding and make little difference to the overall patterns observed.

These topics were generated from the data in an initial reading of a transcript, asking "What are the topics of this utterance?" for each speaker's turn. The transcript was then read a second time, and each utterance was coded with as many topics as fit. This reading revealed that several topics were rarely used, and a third reading was done to check if additional occurrences of those topics had been missed.

Categories of Topics	Code	Description of Code (Keywords)
Teaching	**Gr**	Grouping (groups or pairs, think-pair-share)
	I	Interaction (student involvement, teacher prompting, brainstorming, student feedback, *funneling*, listening, questioning)
	Fo	Format (chalk and talk, lecture, modelling, tell me what you need, student-centered approach, *experiment, review*, choice, scaffolding)
	B	Basis (problem based, skill based, language based, reform, multiple solutions, examples, *metaphor/model*, real world)
	Pa	Pacing (working at the pace of the student, pressure to get through presentation)
	Te	Technology & materials (Smartboard, PowerPoint presentation, notebook, worksheet)
	A	Assessment (student accountability, summative evaluation, monitoring, competition)
	Go	Goals of teaching (*abstract thinking, connection to applications*)
	Ta	*Type of tasks* assigned (*projects, performance tasks*)
Learning	**E**	Emotion (motivation, engagement, anxiety)
	H	How learning happens (memorizing, repetition, activity, representations, creativity)
	W	What is learned (organizational skills, study skills, real world applications, *abstractions*, critical thinking)
	Pr	Prior knowledge and experience
	Co	Specific mathematical concepts (equations, fractions)
Institutional	**Cu**	Curriculum (order of topics, goals)
	Ct	Communication between teachers
	Q	Teacher qualifications and perceptions (generalists, specialists, reputation)
	S	Stratification (difference, weaker groups, enriched kids, the strongest kids)
	Fu	*Furnishings (desks, tables)*
	Pb	*Parents' backgrounds*
	La	Language
	PD	Professional development
	Ti	Timetable, scheduling, "the bell rings"
	CS	Class size
	Po	Power
	Pl	Planning

FIGURE 8.1. Topics used in coding regional transcripts

Visualizations

Visualizations were created to assist in seeing patterns in the topics. Each topic was marked by a column and a (non-unique) color.[1] Turns taken by the researcher are indicated by a column on the far left, which is black when the researcher is speaking. These visualizations were made to be looked at from a distance, so the individual words spoken are not visible; however, the number of words in each utterance can be observed. Codes were applied to utterances as a whole, so there is no indication if the topic changed in mid-utterance.

In the visualizations, it is possible to observe patterns. In some episodes, there is a stronger focus on topics related to teaching, and in others, more focus on topics related to learning. Example visualizations can be found in Appendix 8.2 in connection with the analyses and are discussed in Reid, Savard, Manuel, & Lin (2015).

ANALYSIS: QUEBEC

The first analysis is of the transcript of one focus group session of anglophone teachers, categorized into nine episodes (see Appendix 8.1). The teachers first discuss what exemplary and typical teaching is like, and then select the video about fractions that they will share. The following example discourse from Camilla illustrates how codes were applied:

> You know what I find really interesting [about] the whole *Chalk and Talk* thing? Now that I do have not a *Smartboard*, but I have a *BrightLink*, which is the same idea … you're *showing examples on the board*, so if you have one group doing your example on the board and then you save it and then you can use it for your next group. Use it like, the one that I find that the Chalk and Talk does is, is you're *working at the pace of the student* ….

In this passage, the phrases "Chalk and Talk" and "showing examples on the board" refer to the format of the lesson (Code Fo). The words "Smartboard" and "BrightLink" refer to the technology and materials used in the classroom (Code Te). Finally, "working at the pace of the student" refers to the pacing or timing of the lesson (Code Pa). In this case, the word "group" is not interpreted as referring to grouping or stratification, but rather to a class that the teacher teaches. At this level, it is common for a teacher to teach the same course to more than one class. The teacher is noting that interactive whiteboards like Smartboards and BrightLinks allow teachers to save something that was written in one class so it can be used in another class. This was articulated by Camilla:

> I think the process of writing or copying … even at … the *college level* where … teachers will give … suggestions on how to study and they say, recopy your notes.

[1] Although I used color for ease of analyzing the data, the reader does not need to see color to understand the analysis processes as shading under a code is clearly distinguishable without color.

> *Recopy your notes, as a process of learning*, of memorizing or whatever it is that they need to do and [...] because I teach an *enriched class*, I do not allow them to use a cheat sheet during *tests*.

Here the phrases "college level" and "enriched class" refer to different levels of schooling or streams within a school. They are coded as S, Stratification. The phrase "Recopy your notes, as a process of learning" is coded as referring to how students learn (Code H). The reference to "tests" is coded as A, Assessment. The topic of what students learn is indirectly referenced (presumably they are learning by memorizing what is written in their notes, but as this is not a direct reference it is not coded as W).

Observations

Figure 8.2 shows an overview of the topics discussed. For each episode, the dominant category of topics discussed (teaching, learning, and institutional) is shaded. The table makes visible a pattern in the teachers' discussions. In Episodes 1 and 2, when the teachers are first asked to describe typical lessons and exceptional lessons, they focus first on teaching related topics, especially the format of lessons, student interaction, grouping, and the basis of teaching. The researcher also focusses on these topics in Episode 4 when the teachers ask her what she feels is exceptional. In Episodes 1 and 4, however, the focus shifts as the discus-

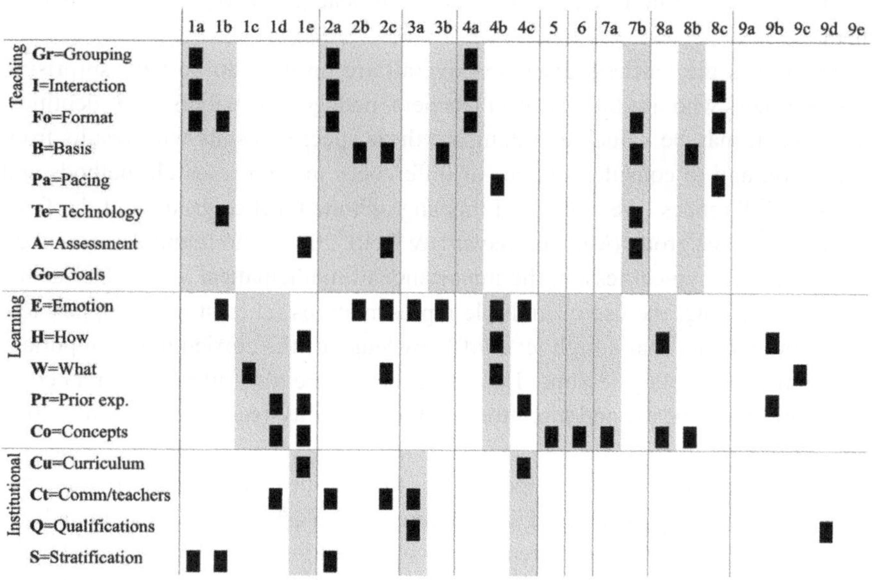

FIGURE 8.2. Overview of Topics Discussed by Quebec Teachers (Shading indicates dominant category in each episode.)

sion goes on, to topics related to learning: *prior knowledge and experience* of students (Pr), *specific mathematical concepts* (Co), *what is learned* (W), and *how it is learned* (H).

In Episodes 5–7, after watching the videos, there is a striking inversion. The first topics the teachers discuss are related to learning, especially, in the case of Video 1, learning *specific concepts* (Co). It is only at the end of the teachers' discussion of Video 1 that they briefly mention topics related to teaching (in sub-episode 7b). The discussion of Video 2 (Episodes 8–9) also begins with the topic of *how learning happens* (H) and learning a *specific concept* (Co), but it then turns to topics related to teaching before returning to topics related to learning after watching the second part of the video.

A further observation is that the topic of reasoning does not occur. This is noticeable primarily in that I, the observer in this case, have a strong interest in reasoning and so would be likely to notice any discussion of it by the teachers. The absence of this topic is an example of a finding that arises out of our methodological awareness that everything said is said by an observer.

Conclusions: Quebec

These results are interesting in several ways. They reflect on the topics teachers themselves find most relevant when describing and reacting to teaching, and on the research methods used and the nature of teachers' pedagogies revealed by them.

The topics the teachers discussed overall are unlikely to be very surprising to researchers who are interested in teachers' pedagogies, beliefs, and identities. However, it may be valuable to compare these specific results with results from elsewhere, and to consider sources of differences in both research methods and regional differences. We compared this anglophone Quebec group and the francophone Quebec group (Manuel, Savard, & Reid, 2014), and found agreement on the format of a typical lesson, the importance of mathematical vocabulary as the basis for teaching, the use of multiple representations (at least in exemplary lessons), and a belief that a high level of knowledge of the curriculum is important in planning exemplary lessons. However, there were also differences related to questioning, synthesis, and attention to student ability (see Manuel et al., 2014 for more details).

From a methodological perspective, it is thought-provoking and important that the topics the teachers discussed were different when asked to describe typical and exceptional lessons, and when reacting to videos of lessons. The teachers did not simply say different things in these two contexts; they focus on different *topics*. This means that a research design that relies on a single way of ascertaining teachers' views of teaching will miss some topics and overemphasize others.

ANALYSIS: ALBERTA

The episodes analyzed here come from the transcripts of the first two focus group sessions. In the first session (Episodes Q1 & O1, see Appendix 8.3), the teachers discuss their teaching in response to questions asked by the researcher, and then planned the video viewing. Between the two sessions, they viewed their tapes separately. They then asked if it were possible, rather than selecting from the edited tapes, to create a composite video from several lessons. In the second session (Episodes O2-I6), they viewed some of the videos together in order to select segments to include in their composite tape. In Episodes Q1, Q2, and Q3, the researcher asked several questions, which are outlined in Figure 8.3.

A non-question intervention is a comment by the researcher in which the researcher makes a suggestion or relates an experience as a contribution to the conversation. An example is the following exchange that occurs between the first and second questions asked in Episode Q2 (between Question 2-1 and Question 2-2):

Maryanne: Yeah, I don't really—I'm thinking mine, just, I'm giving them a problem and I'm having them work on it, checking for understanding and then coming back and kind of going through it. So, correcting misconceptions on, probably, a wider scale. Because if they were more individual, I would help those students as I was walking around, but when it's kind of like seeing the common misconceptions, coming back together as a group and doing more of the teaching then, I guess.

[Beginning of Intervention 1]

Researcher: Yeah, so I think one of the things—because we can also, the other thing we could do is imagine having a couple trajectories, like, so here's one example—'cause everybody kind of starts with this "giving a problem" in some sense. Or giving a task of some kind, that seems to be a common splice. But then you get these different trajectories of the playing out of that, and I think there can be space for that, too.

Brenda: Oh, okay, yeah… Now, um, 'cause we were talking specifically about our algebra one. So is there going—'cause I know you wanted me to do a fraction one. Is there going to be—are we giving them one video, or are we looking at the different—like, are we going to give them three, or…

Maryanne: Or it's not—is it not subject- or topic-related, but just more…

Brenda: Right, or is it just

Maryanne: —the way the lesson …

Brenda: —just pedagogy, yeah.

Episode	Question Number	Question
Q1	1-1	What do you think are three elements or features of your pedagogy that make a difference to learners? Like, if you said, 'Oh, I think doing this is important, and this is important, and this is important.'
Q1	1-2	So is it the math talk that makes a difference, is it working together, is it-is the introduction of problem solving, or is introduction of, sort of, repetition …?
Q1	1-3	Do you think that other teachers are focusing on math talk in their classroom?
Q1	1-4	What are three critical things that the learner in your class must do to be successful? So, what does it take? If for you, it's providing this environment of math talk, working together, problem solving, providing space for repetition, then what is required of the learners?
Q2	2-1	What would you like to exemplify from your teaching [in the composite video they are planning]?
Q2	2-2	Does the topic have an implication for pedagogy?
Q2	2-3	The algebra tiles—what do you think makes the algebra work differently? Than fractions? What is it about that?
Q2	2-4	[Do] you ever need a common denominator [in real life situations]?
Q2	2-5	So how important is making connections to real life to your pedagogy?
Q2	2-6	Would you spend more time filling in the gaps, or more time trying to engage students?
Q2	2-7	What's your strategy for filling in the gaps?
Q2	2-8	So what are the sorts of things [about differentiation] … we have to speak to?
Q3	3-1	What are some of the ... distinctions [in the pedagogy] … that we get from this [video], from the others?
Q3	3-2	What's your regular practice for the class coming to an end?
Q3	3-3	What role does that last thing [an *exit slip*] play for you?
Q3	3-4	Does this [the videos] sort of give an image of the kind of things that happen in your classroom?
Q3	3-5	Is there something that would happen in your classrooms even some of the time that would look quite different than this?
Q3	3-6	What about you, Maryanne? Does everything look like this, or do you have days that look different?

FIGURE 8.3. Questions Asked by the Alberta Researcher in Episodes Q1, Q2, and Q3.

[Question 2-2]

Researcher: Right, well I guess that's an interesting question, in a way, does the topic have an implication for pedagogy?

In these interventions the researcher, who had been a teacher for many years prior to entering academia, enters into the conversation more as a participant than as a researcher. The interventions present a challenge for the data analysis, as the topics in them come primarily from the researcher, not from the participating teachers. One could argue that, because the researcher has an insider perspective on the community of teachers in Alberta, her comments should be included in the analysis, assuming that if there is a shared pedagogy of mathematics teaching in Alberta, it is also shared by the researcher. (For more discussion of insider perspectives, see Chapter 10.) In the analysis here, however, the intervention episodes are not included.

Coding

The codes that emerged through the coding of the Quebec transcript were used as the starting point for the coding of the Alberta transcripts. During the coding process, several key words/indicators were added, and three new topics were added (See Table 8.1). The additional topics required changes to the abbreviations used for some codes. Several codes were used rarely, or not at all. These differences are discussed in the final section comparing the regional analyses.

The following example, from Brenda, illustrates the application of the new codes, as well as some of the old ones.

> Um, well, it depends, just on a unit I like to, um, have some sort of a-an overarching *project* that'll *hit some of the outcomes*, and, um, just have them working through it. A lot of the time it's just something to *keep their focus* and *attention* on something, so they see a big picture, more than every lesson being an isolated, um, *outcome*, that it sort of brings things together a little bit.

In this quote, the word "project" refers to a particular type of *task* that is being discussed, and is coded with the new code, Ta. The phrases "hit some of the outcomes" and "keep their focus" refer to *goals of teaching*, and are coded Go. The word "attention" is related to the *emotional* state of the students and is coded E. Finally, the word "outcome" is used in a technical sense to refer to an objective specified in the curriculum documents, and so is coded as being related to the topic *curriculum*, Cu.

The topic of *furniture* (Fu) was introduced by Brenda, and led to further comments about desks versus tables.

> I think, um, the *projects*, and, um, so the *performance tasks* more are. And I- and I know there's a lot more teachers trying it now than there used to be. And just the, you know, the *working in pairs*, 'cause I always like to have *tables*, and old math teachers don't like *tables* in their classroom, they want *desks*.

However, the topic of furniture did not come up again in later episodes of the Alberta data. Furniture could have been included in the category Te (*technology and materials*), but was kept separate until the comparative analysis. This passage also concerns the topics of *grouping* (Gr), as indicated by the phrase "working in pairs," and *tasks* (Ta) as indicated by "projects" and "performance tasks."

The passages below refer explicitly to *parents' backgrounds*. This led to the introduction of the code Pb.

Brenda:	Our parents don't care. A lot of them—
Researcher:	They care, but they probably don't even have the means to make right—
Brenda:	They don't have the skills, they don't have the time. They are—they had bad experiences, so, you know, they distance themselves, and…

Visualizations and Observations

Visualizations were again created to assist in seeing patterns in the topics. Sample visualizations are included in Appendix 8.4.

Figure 8.4 shows the main codes for each episode. The intervention episodes are omitted, as they are dominated by comments from the researcher, and hence less likely to reveal the teachers' interests. The organization episodes are also omitted, as for the most part, the discussion was not relevant. The responses to Question sets 2 and 3 show the same mixing of topics as in the responses to Questions 1-1 through 14. Depending on the questions, the teachers focused on different topics.

In Figure 8.4, the influence of some of the questions on the teachers' responses is visible. For example, Questions 2-2 through 2-4 are:

2-2. Does the topic have an implication for pedagogy?
2-3. The algebra tiles—what do you think makes the algebra work differently? Than fractions? What is it about that?
2.4 [Do] you ever need a common denominator [in real life situations]?

These questions ask about the role of specific content, and so it is not surprising that the teachers referred to *concepts* (Co) and *what is learned* (W) more often in replying to these questions than in other episodes.

The questions asked in episode Q3 are more open, and this is reflected in the variety of topics discussed. Even questions that suggest topics did not always lead to discussion of these topics. For example, Question 3-2 is:

3-2. What's your regular practice for the class coming to an end?

This question suggests a focus on format, but this does not turn out to be a topic in the teachers' responses.

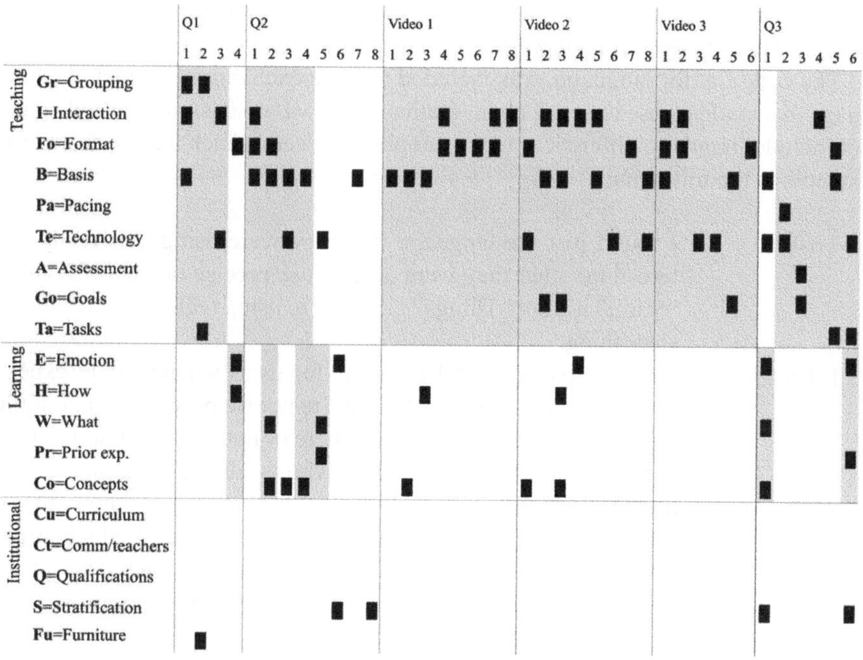

FIGURE 8.4. Main Codes (shaded) in Each Episode for Alberta Teachers (Interventions and organization episodes have been excluded.)

Conclusions: Alberta

Although the researchers' questions and interventions undoubtedly affected the topics the teachers discussed, there is a strong trend, especially when watching the videos and answering the more open questions in Question set 3, to discuss topics related to teaching. Topics related to learning came up in response to some of the questions and while watching Video 2, but usually in addition to teaching related topics rather than being a focus on their own. Institution topics, other than stratification (difference, weaker, groups, enriched kids, etc.) were discussed rarely, and some topics for which codes were created based on the Quebec data, *curriculum* (Cu), *communication between teachers* (Ct), and *teacher qualifications* (Q), were never used in coding the Alberta data.

ANALYSIS: ONTARIO

The data analyzed comes from the transcript of a focus group session in which the teachers are viewing videos together in order to select videos to be shared. The transcript is divided into thirteen episodes, described in Appendix 8.5. The codes that emerged through the coding of the Quebec and Alberta transcripts were used to code the Ontario transcript. Several key words/indicators and new topics were added. These changes are underlined in Figure 8.1.

The code La, for language, was included to code exchanges in which the language that students use to speak about mathematics was discussed. The code does not include linguistic differences (for example, between French and English). An example is the following:

Trista: I found just the language that they were using was really interesting—that they were able to use precise language. It's not "Stuff," it's not "Things".... They're using real words that mean something.

Ellen: Yes, and part of that I guess, is for us as teachers is to expect that in the classroom, and if they're not we prompt them. I don't know if there is an example in there or not but that is something that I do a lot. If there is a child who is talking about something and they're not using that precise language, you know, that's when we prompt them to do that, because that mathematical vocabulary becomes much more important as they... well it is always important but it becomes much more important as they get into high school math, I think.

Professional development (PD) refers to discussion in which the topic is workshops, seminars, and other teacher resources. For example, in this exchange, the teachers are discussing a lesson study program:

Ellen: That was a fantastic program. We got to collaborate with other teachers... took all the feeder schools and the intermediate teachers got together and ... we talked about language in the classroom, we talked about manipulatives ... when the kids came from us to them. It made so much sense and I learned so much because I'm not a math specialist and I know what I do in elementary school and I remember some of my high school math but not a whole lot so I learned a lot about ... where they're going to when they leave here and it was so valuable ...

Jordan: Typical thing—they throw a little bit of money at it and then poof it disappears—there is no kind of sustained....

Ellen: No continuity. [multiple teachers talking over each other]

Exchanges in which school scheduling and timetables are discussed were coded Ti. References to markers of time, such as school bells ringing, were included. For example, the following comment by Ellen was coded as timetable related: "10 minutes a day is 50 minutes, that's a full period a week that's gone thanks to announcements so it's a heck of a lot of instruction time."

When groups of students are discussed as being stronger or weaker, the discussion is coded as S (stratification). However, when the number of students is the topic, the code CS (class size) is used. This is exemplified in a comment by Jordan: "I think I had 31, 31-32. So... they were a pretty busy group, and quite a range."

The code Po (*power*) was introduced to mark passages in which the teachers are discussing issues of control, decision making, and authority in the classroom. For example, here, Jordan explicitly talks about control and decision making:

> If you let them decide where they sit when they come into the room, all the yahoos will wind up in the back-left corner and make your life miserable and set the tone for the year. If you set it, then that says to them, "Well this is about control" and so the cubes are chosen. I don't control, they've got no control. It's the roll of the dice, it's completely random and that's where you sit, and then I'll make adjustments after that if things aren't working out. And generally once, in terms of student seating, once we arrive at a comfortable working relationship and there is a degree of cooperation between members at the table, I will stop adjusting at that point in time because I think the stability is good.

The code Pl marks passages concerning lesson planning specifically, and issues arising from lesson planning, such as flexibility. This is shown in the following comment made by Ellen: "Well lesson planning is an oxymoron in that case. You can plan where you think it might go but you don't end up going in that direction necessarily."

Observations

Figure 8.5 shows an overview of the topics discussed. It makes visible a pattern in the teachers' discussions. A large █ indicates a main focus and a small ■, a minor focus.

There are interesting differences in the topics discussed in response to the video prompts (V1-4) and the other episodes. Several topics, related to teaching and learning, recurred during the discussion of the videos, especially *interaction* (I), *what is learned* (W), *format* (Fo), and *concepts* (Co). These topics were also discussed in episode I4, when the prompt "Do you see similarities in particular things that you do?" invited further reflection on the videos. The only institutional code that occurred often was *stratification* (S), which is often linked to the four codes I, W, Fo, and Co.

In contrast, during the break (B1) and the discussion shortly after (D1), the focus was on *technology* (Te); e.g., the researcher asked, "Do you have a document camera in your room?" At the end of the session, the discussion turned to *tasks* (Ta), as noted by Jordan: "It could be as simple as the wording of the questions this year."

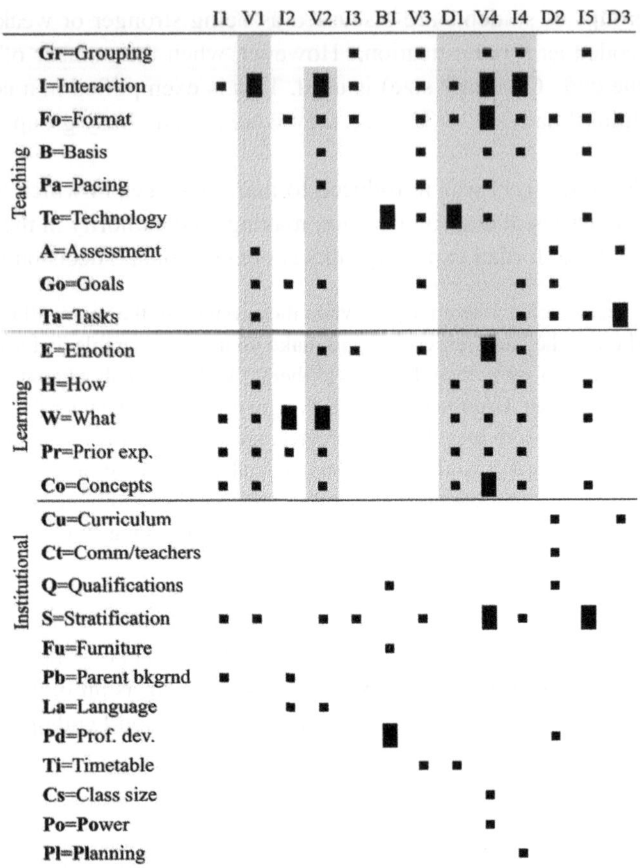

FIGURE 8.5. Main and Minor Focus Codes in Each Episode for Ontario Teachers (Shading indicates episodes where the codes I, W, and Co occur together.)

Conclusions: Ontario

When discussing the videos, the Ontario teachers discussed topics related to teaching and learning, and to the institutional topic of stratification, or ability grouping. At other times, they briefly focused on other teaching topics, such as technology and tasks. Assessment (A), how learning happens (H), prior knowledge and experience (Pr), and most institutional topics were only briefly considered.

ANALYSIS: REGIONAL COMPARISONS

The data considered here are only a small part of the data collected from these teachers, and they are certainly not representative of teachers in their regions.

TABLE 8.1. Researcher Utterances as a Percentage of the Total Number of Utterances

	Quebec A	Alberta A	Ontario A
Total number of utterances	612	896	312
Number of utterances by researcher	144	299	40
% of utterances by researcher	23%	33%	13%

Nevertheless, if there exist regional pedagogies, then some sign of these should be visible in the topics these teachers discussed. Unfortunately, other factors are likely to have had a far stronger influence on the topics they discussed, so any comments about pedagogy must be entirely speculative.

One factor that clearly had an effect on the topics discussed was the choice of questions to ask, and the kinds of interventions made by the researchers in each session. Even a rough count of the number of utterances made by the researcher, compared to the total number of utterances, reveals that the three situations were not comparable. As Table 8.1 shows, the researcher in Quebec was responsible for about twice as many utterances as the researcher in Ontario, and the researcher in Alberta nearly three times as many compared to the total number of utterances. In addition, the researcher's utterances in Alberta especially (but also in Quebec) included statements of opinion, in addition to questions and prompts for further comments from the teachers.

Figure 8.6 shows the coding and number of episodes in which utterances occurred. The data from Quebec are divided into two parts: episodes prior to video viewing (1-4), and episodes during and after video viewing (5-9). As previously noted, there are different patterns of topics in these two phases of the Quebec focus group session. To aid comparisons, the percentage of episodes in which the code occurred is indicated, but these cannot be directly compared as the total number of codes assigned differs. For example, during and after viewing videos in Ontario and Quebec, roughly the same number of episodes were coded, but in Ontario 90 codes were used compared to only 18 in Quebec. This means that *Concepts* was a more significant topic in Quebec than *Grouping* was in Ontario, even though both were discussed in five episodes. Codes that occurred in more than the median number of episodes are in bold to further aid comparison. No effort has been made to weight these codes (within the Ontario data both main (█) and minor (■) focus codes are counted as one occurrence), as the numbers should be considered only roughly indicative. They are intended to suggest areas where a more thorough qualitative analysis is warranted. No direct comparison can be made, as the number of utterances and episodes differs. However, the frequency of some codes in comparison to others may be indicative.

		Region (number of episodes)			
	Code	Quebec Episodes 1–4 (13 episodes)	Quebec Episodes 5–9 (12 episodes)	Alberta (40 episodes)	Ontario (13 episodes)
Teaching	Gr=Grouping	3 (23%)	0	2 (5%)	5 (38%)
	I=Interaction	3 (23%)	1 (8%)	**13 (33%)**	**8 (62%)**
	Fo=Format	**4 (31%)**	**2 (17%)**	**13 (33%)**	**10 (77%)**
	B=Basis	3 (23%)	**2 (17%)**	**17 (43%)**	5 (38%)
	Pa=Pacing	1 (8%)	1 (8%)	1 (3%)	3 (23%)
	Te=Technology	1 (8%)	1 (8%)	**11 (28%)**	5 (38%)
	A=Assessment	2 (15%)	1 (8%)	1 (3%)	3 (23%)
	Go=Goals	0	0	4 (10%)	6 (46%)
Learning	E=Emotion	**7 (54%)**	0	**5 (13%)**	5 (38%)
	H=How	2 (15%)	**2 (17%)**	3 (8%)	5 (38%)
	W=What	3 (23%)	1 (8%)	3 (8%)	**8 (62%)**
	Pr=Prior exp.	3 (23%)	1 (8%)	2 (5%)	**7 (54%)**
	Co=Concepts	2 (15%)	**5 (42%)**	**7 (18%)**	**7 (54%)**
Institutional	Cu=Curriculum	2 (15%)	0	0	2 (15%)
	Ct=Comm/teachers	**4 (31%)**	0	0	1 (8%)
	Q=Qualifications	1 (8%)	1 (8%)	0	2 (15%)
	S=Stratification	3 (23%)	0	**4 (10%)**	**8 (62%)**
	Median	3(23%)	1 (8%)	3 (8%)	5 (38%)

FIGURE 8.6. Frequency of Codes by Episode in the Three Regions of Study. *Note:* Frequencies larger than the median are shown in boldface. Codes not used in all three regions have been omitted. The data from Quebec are divided into two parts, as it was noted that there are different patterns of topics in these two phases.

One topic, *format of teaching* (Fo), occurred more often than most other topics in all three regions. Discussion of *specific mathematical concepts* (e.g., equations, fractions, coded Co) also occurred often in all three regions, but in Quebec chiefly in the episodes during and after video viewing, when it was the main topic of discussion (occurring in 42% of episodes). This suggests that *format of teaching* and *concepts* might be important topics across Canada. It is interesting to note that *assessment*, which is often seen as a central topic in education, was brought up by the teachers in all three focus groups only occasionally.

The topic of *interaction* (including student involvement, teacher prompting, brainstorming, student feedback, funneling, listening, and questioning) occurred in many episodes in Ontario (62%) and Alberta (33%), but not as often in Quebec (23%) and (8%), especially during and after video viewing (occurring in only 1 of 12 episodes). Similarly, *goals of teaching* (including abstract thinking and connection to applications) occurred often in Ontario (46%) and Alberta (10%). In Quebec, *goals of teaching* were discussed so infrequently that the code was not included as a topic in this summary.

Three other topics occurred in more than half the episodes in Ontario, but much less frequently in the other provinces. These are *what is learned* (including organizational skills, study skills, real world applications, abstractions) at 62%, *prior knowledge and experience* at 54%, and *stratification* (mentions of difference, weaker groups, enriched kids, etc.) at 62%.

Two topics were discussed more in Quebec than in Alberta and Ontario. These were *emotion* (including motivation, engagement, and anxiety) (54%) and *communication between teachers* (31%), although communication between teachers was not mentioned during or after the videos were viewed. There was also some discussion of these topics in Ontario (38% and 8%, respectively), but much less than in Quebec. In Alberta, these topics were rarely discussed (13% and 0%, respectively).

Looking at the three broad categories of *teaching*, *learning*, and *institutional* factors, topics related to teaching were discussed more often in Alberta than in Ontario, and more often in Ontario than in Quebec. Topics related to learning were discussed more often in Ontario and Quebec than in Alberta.

Figure 8.7 contains a summary of the patterns in the frequencies of codes. Many topics were discussed only occasionally, and only preliminary data were analyzed; so it is unwise to speculate about these patterns. There do seem to be topics that are of common interest to these teachers, and which may be topics that are important to the pedagogies of teachers across Canada. There are also differences among regions, most markedly between Quebec and the other two regions. Two topics, *emotions* and *communication between teachers*, were discussed more often in Quebec than elsewhere. Two topics, *interaction* and *goals of teaching*, were discussed less in Quebec. This may indicate that the pedagogy of English-speaking teachers in Quebec has a different emphasis than elsewhere in Canada.

Pattern	Codes
Discussed often in all three regions	Fo (format) Co (concepts)
Discussed more often in Ontario and Alberta than in Quebec	I (interaction) Go (goals of teaching)
Discussed more often in Ontario than in the other two regions	W (what is learned) Pr (prior knowledge and experience) S (stratification)
Discussed more often in Quebec than in the other two regions	E (emotions) Ct (communication between teachers)

FIGURE 8.7. Patterns Observed in the Frequencies of Codes

Although it would be interesting to know if these patterns persist in the full data set, including the French speaking teachers and the discussions of lessons from other regions, any conclusions from that data would be similarly tentative, given the small numbers of teachers in the focus groups.

REFERENCES

Manuel, D., Savard, A., & Reid, D. (2014). Observing teachers: The mathematics pedagogy of Quebec francophone and anglophone teachers. Poster. In P. Liljedahl, C. Nicol, S. Oesterle, & D. Allan (Eds.), *Proceedings of the Joint Meeting of PME 38 and PME-NA 36* (Vol. 6, p. 360). Vancouver, Canada: PME.

Reid, D. A., Savard, A., Manuel, D., & Lin, T. W. J. (2015). Quebec anglophone teachers' pedagogies: Observations from an auto-ethnography. In K. Krainer & N. Vondrová (Eds.), *Proceedings of the Ninth Congress of the European Society for Research in Mathematics Education* (pp. 3115–3121). Prague, Czech Republic: Charles University in Prague, Faculty of Education and ERME.

APPENDIX 8.1. QUEBEC EPISODES

TABLE 8.A1. Episodes, First Focus Group Meeting, Anglophone Quebec

Episode	Transcript Line Numbers	Description of Episode
1	5-201	Responses to the question "What do you think a typical class in Quebec in the English system looks like?"
2	203-266	Responses to the question "You're saying wow this is a really exceptional lesson. What would you be looking at?"
3	269-290	Following a pause, reactions to "It's interesting. I don't know if you guys are interested."
4	296-354	Discussion following T's responses to being asked by S, "What do you think T? About exceptional-"
5	373-405	Discussion following viewing of the first part of Video 1
6	409-419	Discussion following viewing of the second part of Video 1
7	423-444	Discussion following viewing of the third part of Video 1
8	449-494	Discussion following viewing of the first part of Video 2
9	499-611	Discussion following viewing of the second part of Video 2

APPENDIX 8.2. ANALYSIS OF EXAMPLE VISUALIZATIONS FROM QUEBEC

Episodes Focused on Teaching

Figure 8.A1 shows the topics discussed at the beginning of Episode 1, when the teachers were asked to describe a "typical class." The focus is on *Grouping* (Gr), *Format* (F), and *Stratification* (S). *Interaction* (I), the *Goals of teaching* (Go), and *How learning occurs* (H) also come up. The focus is on topics related to teaching.

Figure 8.A2 shows the topics discussed at the beginning of Episode 2, when the teachers were asked to discuss what an "exceptional lesson" looks like. Again, the focus is on *Grouping* and *Format* as well as *Interaction*, which are topics related to teaching. As in Episode 1, *Stratification* also comes up later. *Communication between teachers* (Ct) is also mentioned.

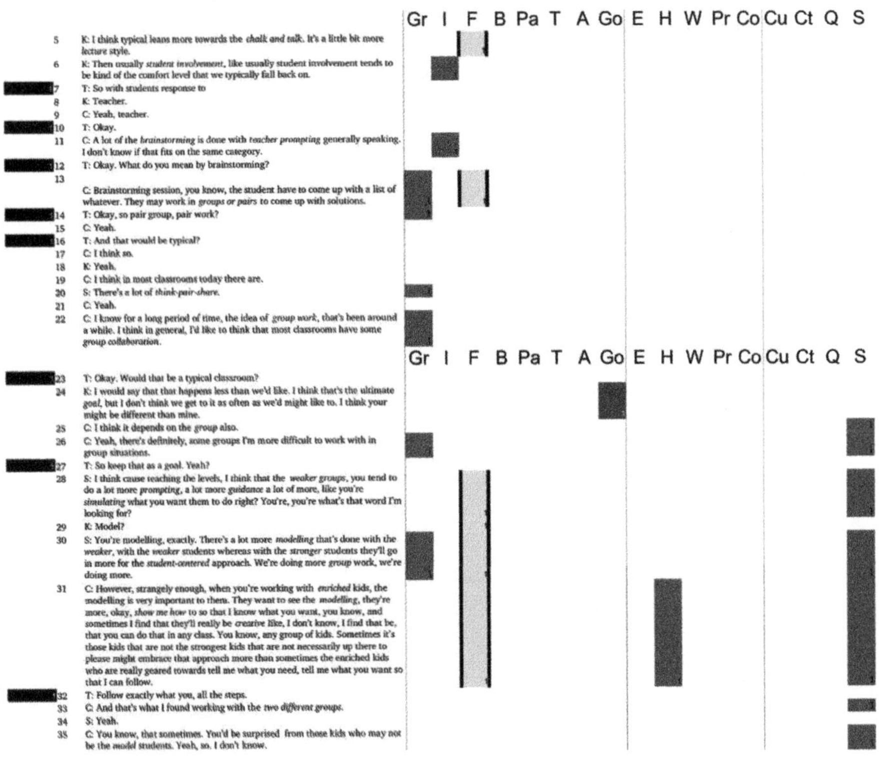

FIGURE 8.A1. Topics Discussed at the Beginning of Episode 1 by Quebec Teachers. Note: Although color is used for ease of analyzing the data, the reader does not need to see color to understand the analysis processes as shading under a code is clearly distinguishable without color.

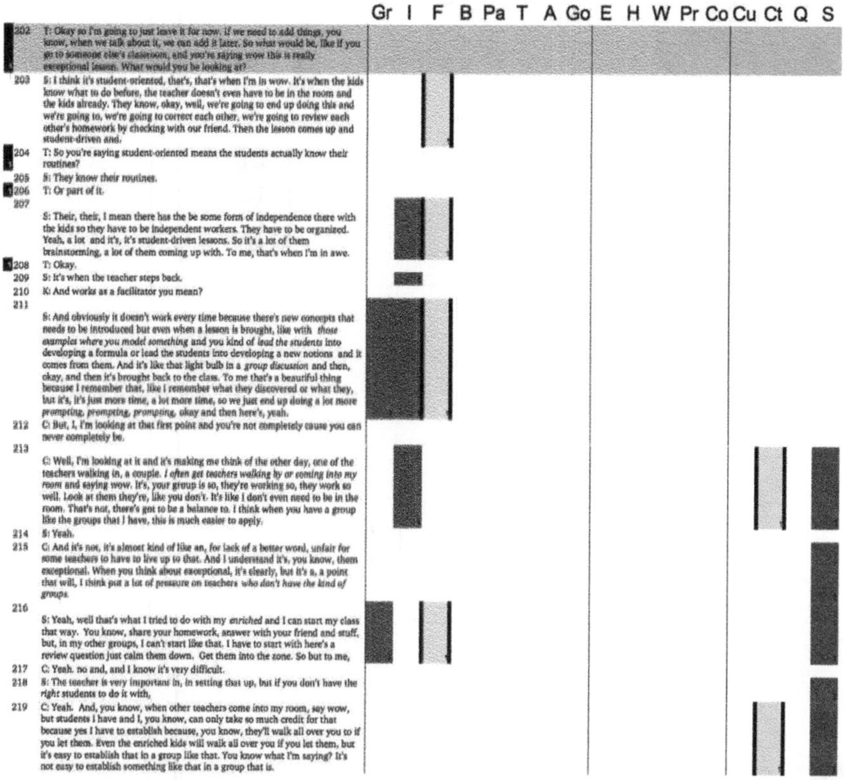

FIGURE 8.A2. Topics Discussed at the Beginning of Episode 2 by Quebec Teachers

These two sub-episodes display a pattern of focusing on topics related to teaching and stratification, with little or no mention of learning related topics. This pattern was observed in five sub-episodes. Of these sub-episodes, four of the five occur in reaction to prompts to describe a typical lesson or an exceptional lesson. This suggests that teaching related topics are the first to come to mind when these teachers describe lessons.

Other topics related to teaching are discussed in sub-episode 1b (focused on *technology use*) and sub-episodes 2b, 2c, 3b, and 8b, focused on the *basis of teaching*.

Episodes Focused on Learning

Figure 8.A3 shows the topics discussed in sub-Episode 1c. The focus shifts away from teaching related topics, although *technology* (T), the *basis of teaching* (B), *assessment* (A), and *format of lessons* (F) are all mentioned. Instead, the focus is on an aspect of learning, specifically *what is learned* (W). The teachers

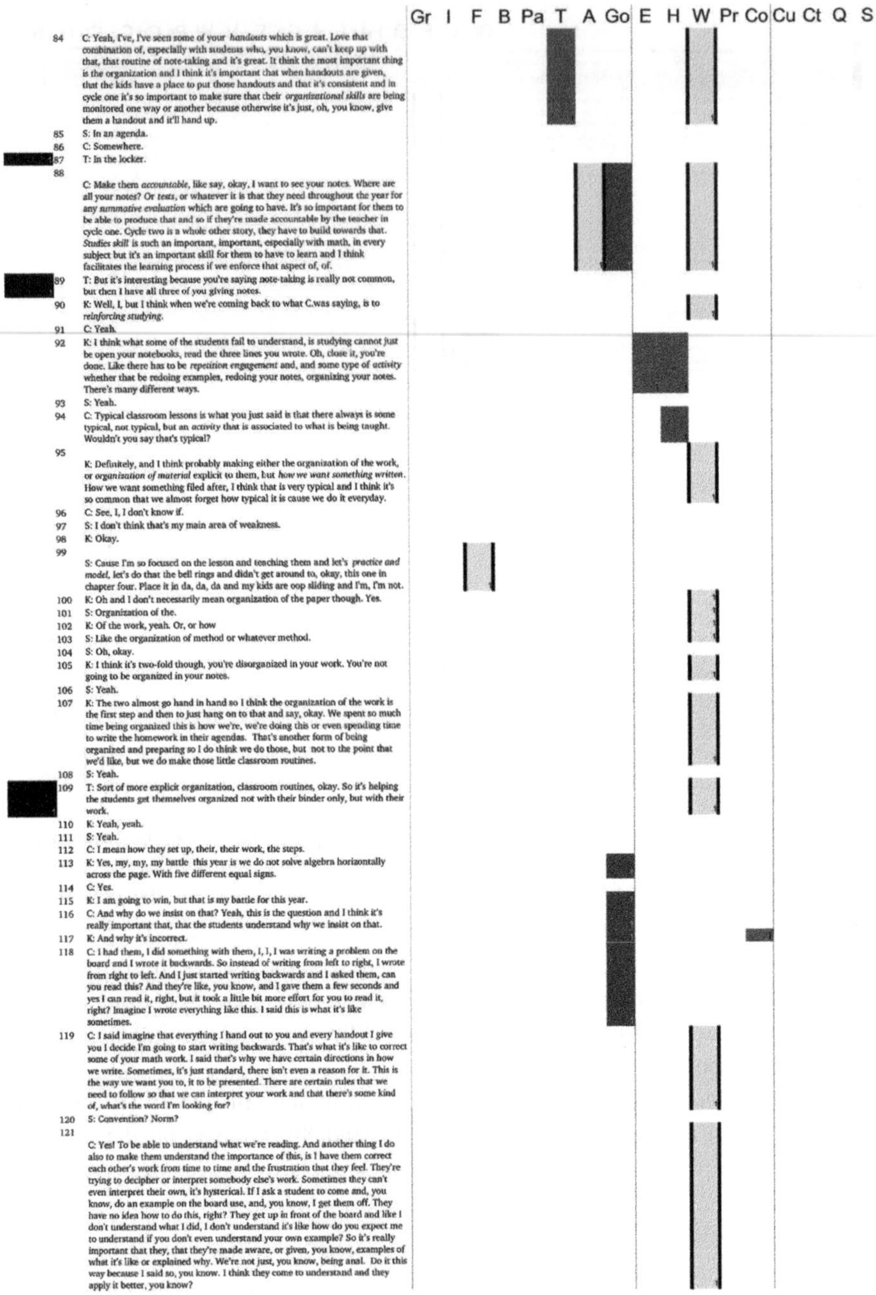

FIGURE 8.A3. Topics Discussed in Sub-Episode 1c by Quebec Teachers, Focused on What is Learned (code W)

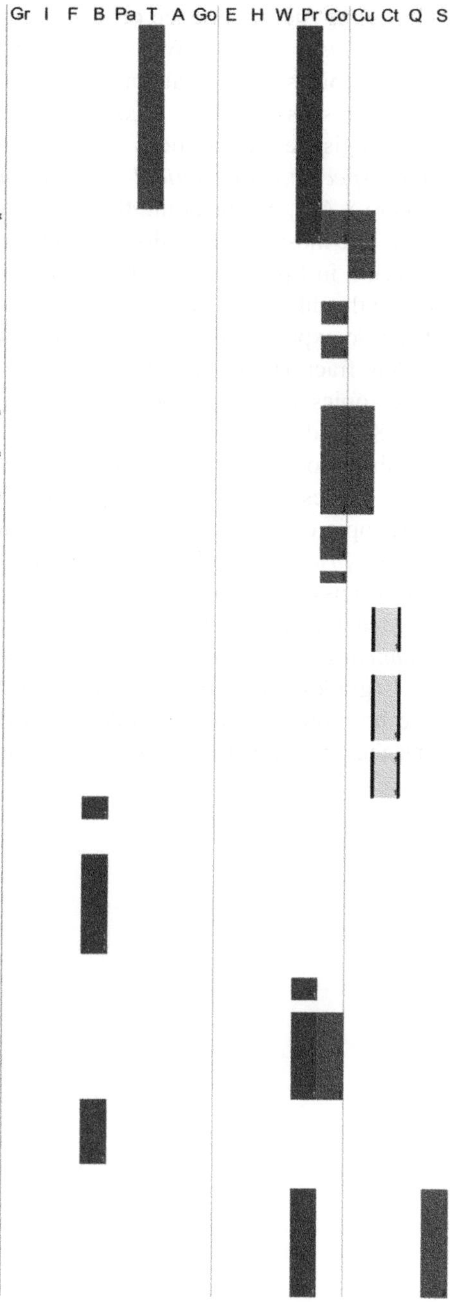

122 S: I think it also comes from a few places where *in elementary there tends to be more one step process* so they're used to doing setting up the operation, equals. So they're used to working that's like, that's what they've been doing for six years. It's just working with Or *workbooks* that give you one line to do your answer. So they're forced to work horizontally unless cause god forbid if they take out a loose leaf, so, right. When it comes time and you give them that space, okay guys, this is and there's a reason for this. Like this is a process, it's like doing a science lab. There's a process to this. You need to identify what I, I, they see it like once they develops in front of them, they see it. But I think there's been, they're so afraid of change and they've been drilled for 6, 7, 8 years to just work horizontally like that's where it all, that's where it started, right? 2+2=4 all horizontal so idea of, you know, pulling out information and then showing your work and working horizontally and because they don't see the bigger picture, right? They don't see balancing equations. So they don't see where this, why we're doing all this work horizontally how it's going to help them later.

123 C: Idea of balancing equations or balancing I find that if, we need to start doing that a lot sooner, I mean coming from the elementary school. The idea of balancing is not introduced at the elementary level. It is, but.

124 T: It is in their curriculum though.

125 C: It is in their curriculum, but it's not. I find that a lot of the teachers will revert back to just

126 S: Do it backwards?

127 C: Yes.

128 S: They don't say you're balancing, they just say well this is the process. Number plus blank equals.

129 C: Right.

130 S: Okay, just do it backwards. They don't, they don't teach what you're actually in fact doing.

131 T: So they're not talking about the equal sign as an equivalence?

132 S: No.

133 T: But it's to find an answer.

134 S: Exactly.

135 K: I find that hard to, to introduce though because really what that hinges upon is your understanding of the zero principle with integers. If you don't have an understanding of integers, that balance really, hangs in the balance. And I really found and having worked with my grade 8s for a couple of years now, I actually inversed the order that I taught things in and did integers and fractions before I ever even ventured to introduce patterning or algebra where in years past I'd done it the other way around because I actually thought they had a better understanding of algebra, they might apply it better to other situations, fractions and integers. I did it backwards this year and did really, let's build our competencies in integers.

136 C: Yes.

137 K: And I found that's actually helping much more with the algebra because that whole concept of balancing, they get the idea of cancelling out because of zero principle.

138 S: Yeah.

139 K: Instead of just randomly plus three, plus three. Well, why?

140 S: Yeah.

141 K: Well, just because you're doing, you're working backwards, okay?

142 C: Backwards, it's like going back to that fish thing, that, that the other grade seven teachers have, you know, they have assured me they were no longer going to use this method I mean, we agreed on that. And again, this is communication right?

143 K: Yeah.

144 S: Yeah.

145 C: One of the reasons, like, you know, you mentioned earlier, oh, well the three of you were saying that it's not practiced, the, the notebooks aren't practiced, but one of the things that the three of us do and the reason why we've maybe moved towards that is because we communicate with other teachers a lot. And, and I think that's something as a school, we need to do more of. Making those, you know, connecting those

146 T: Different cycles, different

147 C: Cycle bridging and even level bridging because.

148 C: You know we say oh cycle bridging. What about level bridging? And this is something that, you know, from one year to another we need to work on more. We're getting better. I know here I think we're better.

149 K: I think so, I think what you were saying that it's vocabulary based, I think all of us have really strived to incorporate literacy in our lessons.

150 T: All of us being you three? Or the school?

151 K: I think us three and then I think it's definitely having.

152 T: It's spreading.

153 K: Trickling effect to others though that we're in contact with because we wouldn't, we wouldn't allow for a biology teacher to teach slang and poor definition for any of the body systems. So why are we allowing to do that, allowing ourselves to do that in math? Like we have proper vocabulary, we need to be using it, we need to be giving examples and I think that's exactly what are, are fish method. Yeah, it's a cute little way of remembering something; however, there really isn't mathematical basis to it. So how are we moving away from that?

154 S: Cross-product , cross multiply, equivalent.

155 K: Yeah.

156 C: And, and there's no reason why it can't be taught that way.

157 C: In grade 7 so. And then to un-teach that. I can't believe how difficult. I think the worse part of how difficult it is to.

158 T: Undo what

159 C: Undo and they, they will put up a fight. They put up a fight, it's unbelievable cause to them it's like, but it works. So why can't we use it? It works, it works and I keep trying to tell them. Yes it works to get your result, but now that we're moving into algebra, it's really important that you balance your, and they don't understand what that means and then they don't understand what I mean by that and it's difficult. I find it very difficult. Maybe there's a better way of explaining this, but I have a difficult time getting them to get on board, you know, so

160 K: I think it has to be done in steps though. I think you have to substitute the slang term for proper vocabulary, model how to do it properly and they connect with the algebra. I think the balance needs to be a three-step process because they're right technically the process will still get them the answer, but maybe they don't have enough basis in the algebra to see the, the bigger picture yet.

161 C: But if they start in grade 7.

162 K: Yes.

163 C: Then by the time they reach grade 8, then we don't have to worry. It's one less thing we have to. There's so much to, you know, have to dealt with in grade 8 and, you know, works. *They're expected to, to have mastered integers and fractions by grade, by the beginning of grade 8*. I'm sorry, that's just not true and I teach the *enriched* kids. I can't imagine, well I can because I have. It's been a while so I feel like I haven't been there in a while, but they need a lot of work in those areas, a lot. And I, even at this point where, you know, were almost in December and I still have some students who are struggling with fractions and integers. You take to grade 9.

FIGURE 8.A4. Topics discussed in sub-episode 1d by Quebec teachers, focused on learning about integers (code Co) and the students' prior knowledge (Pr), with a digression on communication between teachers (Ct)

are discussing the importance of learning good organizational skills in this sub-episode. *What is learned* (W) is also the focus of discussion in sub-episode 4b, where the topic is learning about real world applications of mathematics and episode 9, the discussion of the second part of Video 2.

In sub-episode 1d, the focus is on learning about integers (which is coded as Co, for a *specific mathematical concept*), and the *students' prior knowledge* (Pr, see Figure 8.A4). At one point, the topic shifts to *communication between teachers* (Ct). The most extended discussion in which the focus is on *specific concepts* (Co) occurs in Episodes 5, 6 and 7, in reaction to watching Video 1. There the topics are the difference between negative numbers and subtraction in the context of algebraic expressions (in Episode 5), order of operations (in Episode 6), and equivalent fractions (in sub-episode 7a).

Other topics discussed related to learning include *how learning occurs* (H, sub-episode 8a) and *students' prior knowledge* (Pr, sub-episode 9b).

In sub-episode 1e, the topic of the intended *curriculum* (Cu) came up very strongly. Topics related to learning and *assessment* also came up. Curriculum is also the topic of discussion in sub-episode 4c. Again, topics related to learning (especially *prior knowledge*, Pr) and *assessment* (A) also come up.

The discussion in episode 3 began with reflections on the experience of discussing teaching in the focus group, so the initial topics are *emotions* (code E) and *communication between teachers* (Ct). This shifts into reflections on the status of Cycle 1 (grades 7–8) teachers compared to Cycle 2 (grades 9–11) teachers, who tend to be more specialized (coded Q as this related to *teacher qualifications*). The teachers returned to this topic at the end of the discussion, in sub-episode 9c.

APPENDIX 8.3. ALBERTA EPISODES

TABLE 8.A2. Episodes, Second Focus Group Meeting, Alberta

Episode	Transcript Turn Numbers	Description of Episode
Q1	7-86	Responses to the questions asked by the researcher in the first session
O1	87-110	Organizing tapes to view at the end of the first session
O2	111-119	Organizing the second session
I1-I4	142-317	Non-question interventions by the researcher; these are interspersed with the questions in Q2
Q2	148-326	Responses to the questions asked by the researcher at the beginning of the second session
O3	327-392	Organizing the tapes to watch
V1	393-458	Discussions following viewing of Video 1
V2	459-548	Discussion following viewing of Video 2
V3	549-706	Discussion following viewing of Video 3
I5	614-672	Non-question intervention by the researcher; this occurs during the discussion of Video 3
Q3	707-895	Responses to the questions asked by the researcher at the end of the second session
I6	722-816	Non-question intervention by the researcher; this occurs during Q3

APPENDIX 8.4. ANALYSIS OF EXAMPLE VISUALIZATIONS FROM ALBERTA

Figure 8.A5 shows the topics discussed in response to Question 1-1: "What do you think are three elements or features of your pedagogy that make a difference to learners? Like, if you said, 'Oh, I-I think doing this is important, and this is important, and this is important.'" The discussion is dominated by the topics of *Grouping* (Gr) and *Interaction* (I), with some mention of other teaching related topics, *Format* (Fo) and the *Basis of teaching* (B). The topics of *How learning happens* (H) and *Stratification* (S) are briefly mentioned in this passage by Mary-anne: "Well, with my kids, they're just so weak that … I think what, with my kids, like, a lot of talking, a lot of working together, a lot of repetition …

The responses to Question 1-2 are shown in Figure 8.A6. The question, "So is it the math talk that makes a difference, is it working together, is it … the introduction of problem solving, or is introduction of … repetition…?" touches on the topics of *Grouping*, *Interaction*, and the *Basis of teaching* (Gr, I & B), but the teachers' responses are focused on *Types of Task* and *Furniture* (Ta & Fu).

The responses to Question 1-3: "Do you think that other teachers are focusing on math talk in their classroom?", as shown in Figure 8.7A, focus on *Interaction* and *Technology and materials* (I and Te, respectively). The transition occurs in this comment, made by Andrea, which also touches on *Furniture*.

> I know that some—well, there's a mixed bag maybe. I have teachers in my school who are like, "We're in rows," and it's very, kind of, textbook from however many years ago. You know, "I'll do a Do Now, and then I'll do a lesson, and then you'll do some textbook work," and there might not be any talk at all between kids.

In this case the technology mentioned is a textbook, and the topic of *Furniture* is indirectly referenced by "we're in rows," which could also be seen as referring

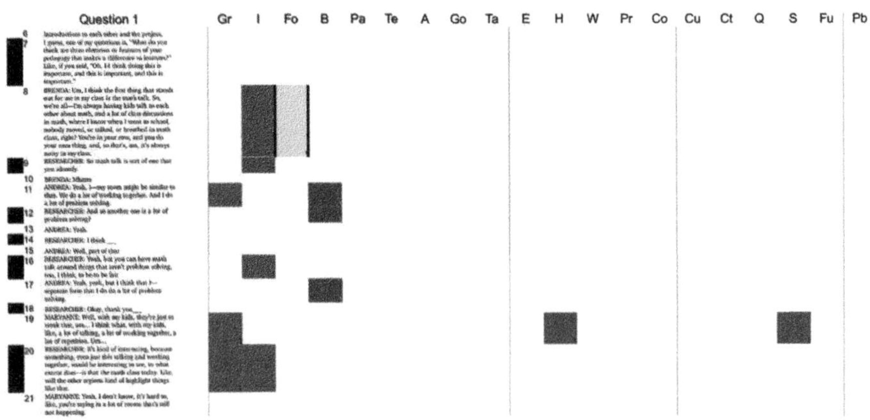

FIGURE 8.A5. Responses to Question 1-1 by Alberta Teachers

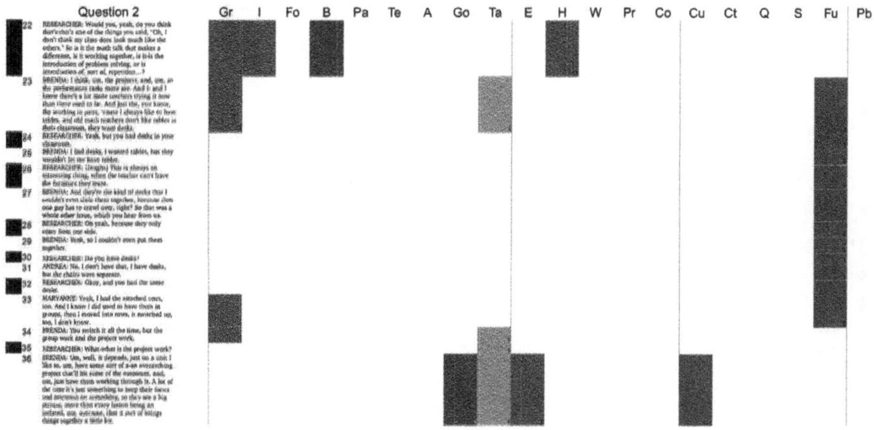

FIGURE 8.A6. Responses to Question 1-2 by Alberta teachers

(more indirectly) to *Grouping*. The brief mention, by the researcher, of *Task* and *How learning happens* seems not to have influenced the focus of the discussion.

Question 1-4 is "What are three critical things that the learner in your class must do to be successful? So, what does it take? If … for you, it's providing this environment of math talk, working together, problem solving, providing space for repetition, then what is required of the learners?" It focusses specifically on *How learning happens* (H), and the topics discussed shift to *How learning happens* and *Emotion* (E), with an interval focused on teaching, specifically *Format* (Fo) and *Goals* (Go) (see Figure 8.A8).

In summary, the discussion prompted by the first set of questions addresses mainly teaching related topics, except when learning is explicitly referenced in Question 1-4. Figure 8.A9 shows all the responses to the first group of questions in a vertically compressed form that makes this pattern more evident.

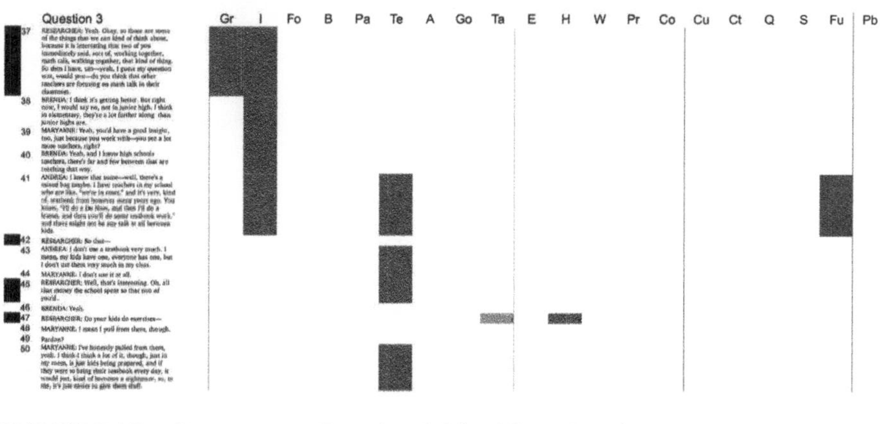

FIGURE 8.A7. Responses to Question 1-3 by Alberta Teachers

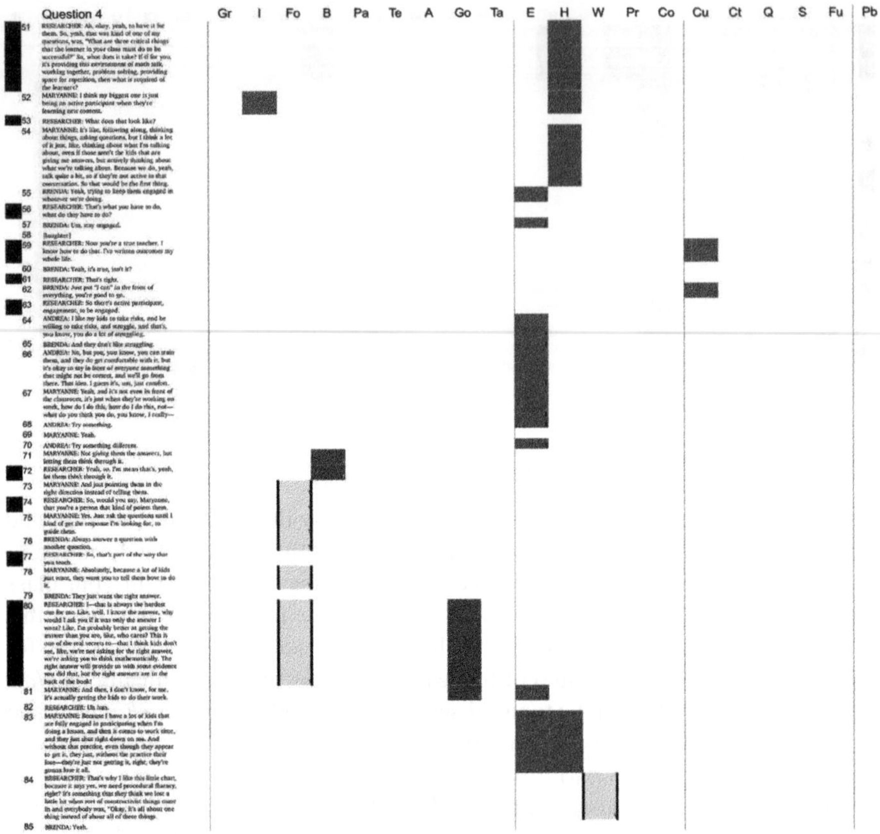

FIGURE 8.A8. Responses to Question 1-4 by Alberta Teachers

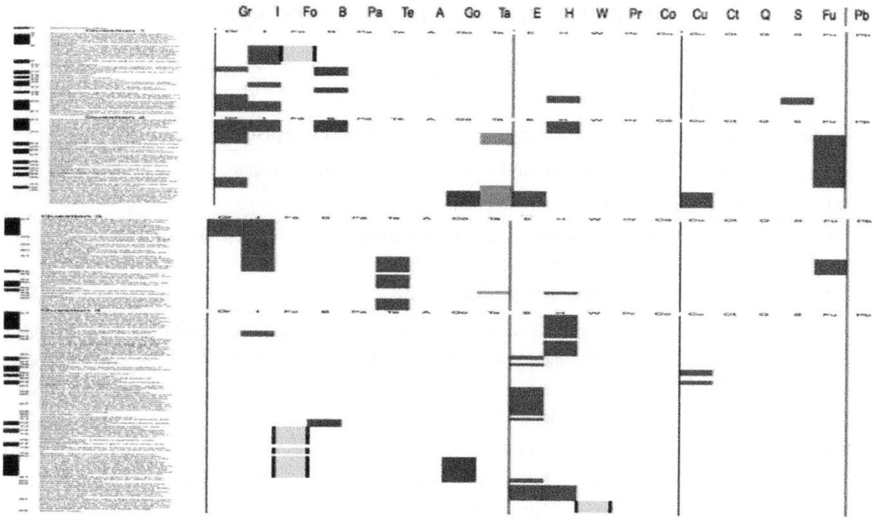

FIGURE 8.A9. Responses to Questions 1-1 through 1-4 by Alberta Teachers

APPENDIX 8.5. ONTARIO EPISODES

TABLE 8.A3. Episodes from the First Focus Group Meeting, Ontario

Episode	Transcript Turn Numbers	Description of Episode
I1	1–18	Introductions and discussion of blurring of faces
V1	20–37	Reactions to Video from School A
I2	38–57	Reactions to I (researcher) comment: "It's interesting that they were able to give so many examples."
V2	59-85	Reactions to Video from School D
I3	86-91	Reactions to I comment: "Having them move from pairs to then working in fours was... is also one strategy of getting rid of that kind of stress level."
B1	92-133	Break: informal discussion continues
V3	135–160	Reactions to Video from School D
D1	161–192	Discussion following a teacher's question "Do you have a document camera in your room?"
V4	194–212	Reactions to Video from School B
I4	213-240	Reactions to I question: "Do you see similarities in particular things that you do?"
D2	242-265	Concluding discussions #1.
I5	266-289	Discussion following I comment: "I'm getting a sense you're not convinced this is necessarily representative."
D3	290-312	Concluding discussions #2

POSTSCRIPT TO PART II

A COMMENTARY ON OUR ANALYSES

In the three chapters of Part II, we presented distinct analyses of the anglophone auto-ethnography data. Our analyses were done separately, though Lisa participated in two of them, and so it is important to take into account our backgrounds described in the introduction to Part II. Here we will compare our analyses, acknowledge similarities and differences that arise from these varied perspectives, and then reflect on influences our perspectives might have had on the analyses.

COMPARING OUR ANALYSES

In Chapter 6, Richard and Lisa were inspired by teachers' own recognitions of commonalities in pedagogy to seek out other utterances that showed similarly how teachers make sense of their pedagogy discursively. In this chapter, they searched the transcripts for a specific type of utterance, namely, one where teachers came to some agreement about their shared pedagogical practices. The teachers had not been tasked with looking for sameness; they were simply asked to identify videos that were somehow representative of mathematics teaching in their province. In the course of these discussions, they indicated and negotiated samenesses, which then became available for analysis. Other parts of the transcripts may have contained interesting insights but were not included in the analysis. In Chapters 7 and 8, the authors examined each regional transcript as a whole. In Chapter 7, Lisa

Researching Pedagogy and Practice with Canadian Mathematics Teachers, pages 187–191.

and Ellen's analysis utilized a thematic approach, whereas in Chapter 8, David's analysis looked at transcripts to code the general topic of utterances. We first compare the two similar approaches in Chapters 7 and 8, and then examine the emerging ideas from Chapter 6 in relation to this analysis.

When examining Chapters 7 and 8, we note that many of David's categories align with the broad themes of Ellen and Lisa's analysis. However, the distinctions made in the two analyses are not the same. For example, two themes that Lisa and Ellen identified (teacher as facilitator and classroom community) are related to two of David's categories (format and interaction) but orthogonally, so that David's categories can be seen as two aspects of both of Lisa and Ellen's themes, and vice versa. At the same time, there are themes that make more precise distinctions than David's categories. The theme of engagement, for example, is subsumed in David's general category of emotion, and so the emphasis on engagement in Alberta is not visible in David's analysis. David's codes also highlight some unique ideas that get lost in the thematic coding. For example, David's category of pacing (teachers working at the pace of the student, and pressure to get through presentation) was identified by teachers in all three areas but was not identified as a theme by Lisa and Ellen. Rather, Lisa and Ellen focused more on how teachers talked about strategies for developing concepts than the concepts themselves, aligning with their approach to research and again reinforcing the idea that everything said is said by an observer.

Examining Chapter 6, we see that many of the themes in Chapter 7 are also present, such as considerations of vocabulary, the importance of engaged learners, and the challenges of covering curriculum. However, there are also variations in the analyses that point out different aspects of pedagogy. For example, the idea of allowing students to struggle within the theme of teacher as facilitator is highlighted in Chapter 6. This idea aligns with some of the pedagogical practices pointed out in Chapter 7, such as teacher as facilitator. However, what emerges in the discursive analysis done by Richard and Lisa is a deeper understanding of the complexities for both teachers and their students relating to this pedagogical approach. Chapter 6 shows how the "struggle" is fraught with tension for students who may feel anxious and uncertain. Teachers equally experience emotional responses to the struggle, questioning when to provide guidance or when to step back and allow the thinking to continue. The analysis in Chapter 7 does not capture this emotional component, while it is captured by David's topic of emotions in Chapter 8. However, David's coding reveals that these emotions occur, but does not give insight into how they are experienced by teachers and students. The discursive analysis in Chapter 6 is able to reveal pedagogical insights in a more holistic way, yet it does not capture the breadth of themes nor the frequency of topics as do the thematic and topical analyses. Although this is just one example, we see that all three approaches to analysis reveal similar ideas about pedagogy, yet each approach also reveals unique aspects of the themes that the others do not.

INFLUENCES ON THE ANALYSES

In these chapters, we have provided three different ways of exploring the same data. We have also noted that, as observers, we come from different places and life histories. Just as we looked for similarities and differences between the three focus groups discussed in the chapters of this section, here we look at similarities and differences in our analyses of the transcripts. These three different lenses allow us to look at pedagogy in different ways.

Given that Lisa and Richard have a common interest in language and culture as factors in learning, it is not surprising that they chose a discursive lens for their analysis in Chapter 6. Lisa is particularly interested in how teachers engage children in mathematical learning experiences that are rooted in cultural ways of knowing, being and doing, and advocates for asset-based approaches to teaching mathematics, seeing children's strengths as a starting point for learning. In her own work, this is intimately tied with an understanding and valuing of the home language of children. Lisa and Richard have discussed these ideas often in relation to work they have been doing separately, particularly with Indigenous youth. Richard's interest in students' languages is consistent with his application of discursive frameworks that avoid making assumptions about what participants are "really" thinking. The idea is to work at the level of the utterance, producing detailed analysis of short stretches of interaction.

Lisa's and Ellen's themes in Chapter 7 reflect their common experience as teachers in Nova Scotia. Lisa's work as a teacher in a Mi'kmaw First Nation community school and Ellen's in rural Nova Scotia give them a common experience teaching in schools removed from the urban centers where educational policies and priorities are largely determined. This may have made them especially sensitive to themes such as differentiation and community building in classrooms. Furthermore, Lisa and Ellen continue to spend considerable time working alongside classroom teachers in both their research and teaching capacities, so their daily work focuses on aspects of classroom pedagogy and relational teaching. This experience also had influence on their analysis. They recognize that it may be the case that what they see are the things they value and elevate in their own work with teachers.

David's approach might seem to be more "objective" than the others. In his analysis, he did not infer the teachers' attitudes towards a topic, but only noted that they discussed it (similar to Richard and Lisa). However, his observations are just as much made by an observer. David has more than a decade of experience as a teacher educator in rural Nova Scotia though he never taught in rural schools, and some of the overlap between the topics he noticed and Lisa's and Ellen's themes may reflect this experience. His research on emotions in teaching (e.g., Brown & Reid, 2006), and on the emergence of disparity among students in classrooms (e.g., Knipping, Reid, & Straehler-Pohl, 2015) may account for his observation of the topics of emotions and stratification. Pacing, a topic David observed but the others did not, has been identified as important in his emergence of disparity research. As noted in Chapter 7, David also has a strong interest in proving and reasoning and so would be likely to notice any discussion of it by the teachers. The absence of this topic in the observations of an observer sensitive to it is an interesting finding.

AREAS FOR FURTHER RESEARCH

Taking these analyses together suggests areas for future research. Lisa and Ellen identified *differentiation* as a theme that was discussed more in Ontario than in the other regions. David's category of *stratification*, however, which seems closely related, marks several episodes in the discussions by the Quebec focus group, but only those discussions that occurred before they watched the videos. A closer look at the nature of those discussions and the ones that occurred in Ontario seems warranted. Similarly, it would be interesting to compare episodes in the theme *student engagement* within the larger category of *emotions*, to see if the discussions in Quebec were broader than those in Alberta.

Although our analyses enabled us to identify some differences among the different regions, they do not enable us to develop explanations for regional differences in achievement. Indeed, we also identified many similarities: teachers in all regions appear to be thinking about the format of their lessons, the concepts they teach, and the nature of the interactions occurring in their classroom, which perhaps reflects an effort to find a balance between procedural and conceptual knowledge (such as in the discussion about student struggle, for example). The differences we observed may be indicative of regionally different concerns among mathematics teachers, but whether these translate into differences in teaching or student achievement remains to be investigated.

We conclude that additional data would be needed to determine whether or not the similarities and differences among the groups are explanatory. We must also recognize that there may be other contributing factors, such as funding, teacher preparation programs, time in class, and so on, that would need to be explored. Finally, as previously mentioned, these findings are situated in the context of the research study, which did not involve any classroom observation. Our analyses are of teachers discussing videos of each other's practice, not analyses of classroom practice itself. Of course, our position that analysis is a kind of relation between observer and observed applies just as much to analyses of classroom practice. Hence, although it would be interesting to compare our findings with similar analyses of observations of classroom practice, such a comparison would not guarantee a higher level of objectivity.

REFERENCES

Brown, L., & Reid, D. (2006). Embodied cognition: Somatic markers, purposes and emotional orientations. *Educational Studies in Mathematics*, *63*(2), 179–192.

Knipping, C., Reid, D., & Straehler-Pohl, H. (2015). Establishing mathematics classroom culture: Concealing and revealing the rules of the game. In U. Gellert, J. Gimenez Rodriguez, C. Hahn, & S. Kafoussi, (Eds.), *Educational paths to mathematics: A C.I.E.A.E.M. Sourcebook* (pp. 67–96). Cham, Switzerland: Springer.

Reid, D. (1996). Enactivism as a methodology. In L. Puig & A Gutiérrez (Eds.), *Proceedings of the Twentieth Annual Conference of the International Group for the Psychology of Mathematics Education*, (Vol. 4, pp. 203–210). Valencia, Spain: PME.

PART III

INTRODUCTION TO PART III

Reflecting on Observing Teachers

Part III of this book contains two chapters and includes general reflections about the Observing Teachers research project. These reflections were written not only by researchers and research assistants, but also by two of the teachers involved. This allows us to present the voices of all the different groups involved in the project.

Chapter 9 is a reflection from two anglophone teachers from Quebec who participated in the research project. It is based on interviews initiated by Terry Lin, who was a research assistant in the McGill team. The team of researchers proposed questions about the impact of the Observing Teachers project on the teachers' teaching in a Google Doc, and the participants, without any specifications from the research team, began a written conversation together in the document, commenting and elaborating on their opinions. In this chapter we present some of their conversation, focusing on how this project enabled the participants to notice important aspects about their teaching and how it changed their identity as teachers. The authors of this chapter kept intact most of the written conversation to show the richness of the content.

Chapter 10 includes reflections from the research team. Because this project was a large one, involving researchers from different regions in Canada with different lenses on pedagogy and on research, some members reflected on and at-

Researching Pedagogy and Practice with Canadian Mathematics Teachers, pages 193–194.

tempted to theorize many of the challenges that arose during the project. Some of the challenges include the teacher-researcher-learner relationships; in other words, how the researcher positioned him or herself during the focus group (as a participant, as an expert, as an insider, or as an outsider), as well as dealing with unexpected events while having to maintain the ethical considerations promised to the participants and attempting to collect data at the same time. This chapter also includes reflections from researchers on what it means to do research.

Despite the many challenges that occurred, this project appeared to be a professional development experience that permitted the members (participants and members of the research team) to grow in their fields. This part of the book highlights some of this growth. While reading it, you, the reader, may wish to reflect on the following questions:

- What have you have learned about the researchers, the participants, and teaching and learning mathematics in Canada?
- What have you learned about the complexities of research and the ways in which research relationships, dilemmas, and ethical decisions reach far beyond the research design and ethics submission?

CHAPTER 9

LISTENING TO THE TEACHERS' PERSPECTIVES

Dialogue on Research Participation and Growth

Terry Wan Jung Lin, Annie Savard, Claudia Anzovino, Kathryn Sides Blanchard, and Dominic Manuel

This chapter was co-written by three researchers and two teachers. The teachers, who wished to be part of the writing process, discussed their participation in the research project through contributions to an online document. They felt that participating in the project was impactful professional development for them, which they wanted to acknowledge by reflecting deeper through a written dialogue. This writing provides insights into teachers' participation in a research project from their point of view.

CONTEXT

Research in education has continuously indicated the importance of teacher professional development to improve the quality of instruction in the classroom. Too often, teacher participation in research projects is not identified as a professional learning opportunity in the literature (Goos, 2008).

This chapter presents the voices of two teachers and co-authors of this article, Claudia and Kathryn, who have been teaching in an anglophone school in Quebec for at least seven years, as well as the voices of the research team that accompa-

Researching Pedagogy and Practice with Canadian Mathematics Teachers, pages 195–211.

nied them throughout this research project. The research team consisted of Annie, the principal investigator, Terry, a research assistant for the first phase of the project, and Dominic, a research assistant during the second phase of the project.[1] The idea for this chapter began when one of the teacher participants mentioned that she and her colleague believed that their participation in this research project was "the best PD (professional development)" they have received in their careers as teachers. Terry was inspired by this comment and wanted to look further at teachers' reflections after the data collection process was completed.

Terry was also a former colleague of Claudia and Kathryn. This prior relationship helped to create a safe environment that was built on trust and respect. Terry facilitated the discussion during the first phase of the project. Annie and Dominic did not know the two teachers prior to the project and were considered the "outsiders" of the group, but facilitated the discussion during the second phase of the project.

This chapter was written with the teachers because they wanted to participate in the writing process on the research project after data collection. To facilitate this, Terry and Annie posted some questions on a Google doc, and the teachers responded to them individually. The questions were:

- What moments in this project were important or meaningful to you?
- What did you notice about other teachers' teaching?
- Why was your participation in this project an important moment for your professional growth?

The two teachers read each other's comments and responded, in writing, to each other as they thought it was very helpful for them. Annie, Terry, and the two participating teachers continued the conversation with online meetings. During those meetings, they discussed the structure of this chapter. Terry, Annie, and Dominic thought that the teachers' dialogues were an important contribution to this chapter. Terry began a thematic analysis of the dialogue, which was continued by Annie and Dominic. The main themes are presented in the next section.

The goal of this chapter is to describe how the teachers perceived the Observing Teachers research project and the impact the project had on them as teachers. The questions posed by Terry, Annie and Dominic that focused Claudia and Kathryn's written conversations are as follows:

- What elements of the research project did the teachers find useful for their learning?
- What conditions were necessary to transform these elements into opportunities for teacher learning?

[1] The first phase of the project included teachers viewing videos of their own teaching; the second phase included teachers viewing videos of teachers from other provinces.

- How did the teachers perceive the impact of their learning in their current work?

The videos shown in this research project provided a change of pace from the videos that teachers often view. More specifically, the teachers did not come to the project thinking that they would be watching videos that accompany published books, which show expert teachers demonstrating certain practices. Rather, in positioning themselves as equals to the other teachers in the lessons that they observed, they seemed receptive to noticing similarities and differences in teaching methodologies, and disposed to reflect critically on the approaches that they were seeing.

TEACHERS' PROFESSIONAL VISION

Ball and Cohen (1999) argue that teachers must be able to notice aspects of pedagogy and interpret what is happening in their classrooms in new ways to be able to adapt their teaching in the midst of instruction. Sherin (2001) observed this ability as teachers developing their *professional vision* through their participation in video clubs. Professional vision refers to a way of describing the ability that members of a professional group share for interpreting phenomena central to their work (Goodwin, 1994). In Goodwin's words, professional vision involves "ways of seeing and understanding events that are answerable to the distinctive interests of a particular social group" (p. 606). This entails the ability to notice features of practices that are valued by a group. and Sherin (2008) argue that teachers may need to further develop their abilities to notice, particularly in reform teaching.

Video clubs consist of groups of teachers who meet to watch and discuss excerpts of video records of their instruction (van Es & Sherin, 2008). We used the concept of video clubs as it relates to the experience teachers had in this research project. Although video clubs are a context that require further study to fully understand the nature of teacher learning, they appear to show promise in supporting teachers in learning to notice, as videos are able to capture much of the richness of classroom interactions and can prompt teachers to reflect on these interactions (van Es & Sherin, 2008).

van Es and Sherin (2002) developed a *Learning to Notice Framework* that consists of three main aspects: identifying what is important in a teaching situation; using what one knows about the context to reason about a situation; and making connections between specific events and broader principles of teaching and learning. The first aspect represents being able to identify what is important in situations, which Fredriksen (1992) refers to as *call-outs*, and to use *check points* to assess the progress of lessons. Goodwin (1994) used the term *highlighting* as the ability to bring certain elements to the fore. The second aspect of the framework aligns itself with previous research, which suggests that individuals are more able to make sense of situations they encounter in domains as they gain experience in them (Brown, Collins, & Duguid, 1989; Chi, Glaser, & Farr, 1988; Lesgold et al., 1988). According to van Es and Sherin (2008), "for teachers, this means using knowledge of the subject matter, knowledge of how students think about the subject matter, as well as knowledge of their local context to reason about events as they unfold" (p. 246). This type of reasoning is usually influenced by teachers' knowledge and beliefs (Schoenfeld, 1998). The third aspect of the framework is influenced by previous studies, which suggest that experts consider problems in terms of the concepts and principles that specific situations represent (Glaser & Chi, 1988; Larkin & Simon, 1987). Shulman (1996) used the term *extrapolating* from the specific to the general to describe this aspect. According to van Es and Sherin (2002), extrapolating enables teachers to observe a situation and see it as an instance of a principle of teaching instead of as isolated events.

The following dialogue between the two teachers, Claudia and Kathryn, highlights their professional vision about being a participant in this research project and speaks to their professional identity. These dialogues were based on what they noticed from observing video-recorded lessons of themselves and their colleagues, as well as teachers from other regions. The dialogues these two teachers wrote were based on elements from the research project that greatly impacted them: their practices, their identity as a teacher, and their professional growth. As professional growth, participating in the research project impacted their identity as a teacher, and they shifted from teacher-participants to teacher-researchers. Both the researchers and the teachers found that the content of the conversations was very rich. The following presents a descriptive analysis of the two participating teachers' reflections. For each main theme, the researchers present a description of the key elements discussed by the participating teachers, and then present the discussion between then.

Feeling the Impact of Participating in Research on Teaching Practice and Teacher Identity

As part of the project, the teacher participants were asked to engage in discussion about their vision of what a *typical* and an *exemplary* lesson would look like. Characterizing typical and exemplary lessons was at the heart of this research project because teachers had to select a video for each type of lesson to share with the other regions. Claudia and Kathryn reported that these two terms were abstract to them. The idea of viewing lessons and deciding whether they were typical or exemplary made them feel uncomfortable. At the same time, this seemed to make them think more deeply about their classroom practices, such as the types of questions to ask students, and how to keep their instructional goals in mind so their lessons would be more exemplary.

Kathryn: I felt really uncomfortable at first deciding whether videos would be categorized as exceptional or typical. Typical, even though not a negative term, felt to me like we were ranking the lessons as lesser. It took a while for me to wrap my mind around the fact that typical meant that it would be a lesson we could expect to see in any classroom on any given day. Exceptional lessons were truly masterful in the delivery of the material, engagement of the students or the supporting materials used.

Claudia: I felt the same way. I believe it is our natural instinct to strive for exceptional lessons when we are placed in a comparative situation. With collaboration there is a greater likelihood that exceptional lessons will be typical. This was interesting for me since my understanding of a typical lesson versus an exceptional one was abstract at best and I had no tools for measurement. I realized after watching all the videos including my own, that there was no clear answer but only our individual perception of what we chose to focus on: content, classroom management (class dynamics), teacher/student engagement, use of technology, manipulatives (tools), peer collaboration, or vocabulary.

As shown in the quote that follows, Claudia noted how exemplary lessons can go in multiple directions. This shows the complexity of teaching in terms of making "on the spot" decisions to attain the goals of the lesson.

Claudia: At the end, even though I was left with a somewhat clearer idea of what an exceptional lesson should consist of, it is still an abstract and vague idea. However, it is that vagueness that pushes me to strive to achieve it every day. Before the video project, I never thought of it like this. I never asked myself whether I found my lesson to be typical or exceptional. Most of the time, as long as I and my students stayed on task, accomplish the goals I set out to achieve, my lesson was successful. Now I strive for more. When I feel motivated (because some days I just don't have the energy and a typical lesson will have to do, and that's okay too), I look for teachable moments; where beginning with asking the right questions is key to grasping where they're coming from and building on their current knowledge. This knowledge will set up the goals of the lesson. This is risky since the lesson can go in many different directions and one needs to be prepared to make mistakes and have lessons that go sideways. This approach works really well when introducing a new concept (save the formal PowerPoint lesson for the next day). I remember often feeling that my best lessons were the ones I winged and could never quite figure out why. Now I get it. When I wing it the students automatically influence the direction of the lesson making it more authentic. These are also the lessons that can never be replicated.

At first, Claudia and Kathryn felt nervous and vulnerable to be watching videos of themselves teach in front of other colleagues. But the safe environment created by knowing each other gave the participants comfort as they felt they could receive constructive feedback about their teaching. This is shown in the following dialogue.

Kathryn: I was initially really nervous about watching videos of myself, but I soon discovered that watching videos of myself with my colleagues was an empowering experience but also required a lot of vulnerability.

Claudia: Feeling vulnerable is essentially what I felt as well. However, I also felt safe since those watching with me were people I trusted. I felt that they would use what they see to provide me with constructive feedback. This is what empowering should feel like and this can only happen when you have a feeling of trust with the people you work with. Vulnerability comes with trusting those you collaborate with. What do you think about the trust factor?

The group dynamic during the focus groups seemed to have led to deep and rich discussions about teaching mathematics. In addition, according to the participants, noticing elements of their teaching made them realize common elements in lessons and elements that were important to them, such as the collaboration between the teacher and students. It led them to rethink their role as a teacher. In fact, noticing permitted them to shift to another important reflection of their teaching: being a facilitator by planning better questions to ask students instead of creating a good lesson.

Kathryn: I think the pre-existing relationships our group had as colleagues, created a safe space for this vulnerability and then permitted a very open dialogue of questioning and commenting. It was still hard to initially watch myself because I think that we are always our own harshest critics. I was surprised that while I spend time looking at details of my perceived weaknesses the rest of our group focused on the connections I made with my students and engagement of the class.

Claudia: I remember getting feedback about myself that I wasn't consciously thinking about while I was teaching. One of the elements of my teaching that my peers pointed out was the use of rich math vocabulary which I was using, defining and repeating throughout my lesson but to me this wasn't deliberate for this lesson but instead just a naturally occurring happening in my classroom. I am now very conscious of this and use it deliberately in my teaching.

Kathryn: It sounds like a strange realization, but I realized that my role as a teacher was beginning to shift to facilitator and that my classroom was not all about me. For the longest time I thought my preparation, pacing, planning was what drove the lesson, but I think the video drew my attention to how my questions were letting the students drive the lesson. This created a huge shift in my teaching and planning and now I plan great questions instead on spending that time on material.

Claudia: I like the word facilitator. I think this word redefined the role of the teacher. It is important for students to work collaboratively and build trust as well. They, too, need to feel safe to share their vulnerabilities with their peers in order to learn.

Kathryn: I remember that from watching your [Claudia's] class, the sense of community and respect truly gave every student a voice in building the collective understanding. You also teach with such enthusiasm that your students echo back that same level of excitement for learning.

Claudia: I find collaboration was a common practice in many of our videos. You too had a sense of community and trust in your classroom. Your students were free to ask questions and were given time to discuss solutions with their peers. This was continuously built into your lessons setting up a community of openness and trust.

Kathryn: You're right, I do build in wait time for my students to ensure that they have time to reflect on their responses. I think, and we've discussed this before, that we teach instilling the values that are important to us, in addition to the curriculum. Respect, community and precision of language are ever present in both our classrooms because they are values that are important to us. I think watching our videos made our values more evident to others and more salient to myself.

Claudia: I think our value of the importance for collaboration is also evident in our teaching.

Kathryn: Absolutely! I think watching these videos exposed who I am as a person and by default that is what I bring into my classroom every day. I used to think my identity as a teacher was somewhat separate from myself but after watching my videos, I realized that I am teaching what I believe in and my colleagues helped identify these values.

Claudia: Wow! I think this is exactly why this project was so transformative. You just put in words what we experienced and what we had a difficult time putting in words.

The two teachers shared their vision and their thoughts about observing other teachers' practices. It seemed that observing others led them to reflect about their own practices and identity as a teacher. Kathryn and Claudia were able to notice elements that were important to them and felt more attached to lessons that matched their values. Some of those lessons were the ones that were more collaborative between the teacher and the students, and those in which the direction of the learning could be followed. In contrast, teacher-directed lessons and those that used technology in a direct way did not attract them. In their reflection, they questioned if their values made them biased while watching the lessons taught by other teachers.

Kathryn: I remember really relating to some of the videos we watched and feeling really connected to them. I felt so inspired by the videos where there were whole class discussions and the class working to construct their understanding. I also remember feeling very disconnected from videos where a lot of technology was being used. I wonder if I was connecting to videos where the teachers displayed similar values to my own.

Claudia: I agree. Being that collaboration is a personal value we share, then it is no surprise that we both did not connect with videos that were primarily teacher directed. Often and unfortunately, technology is used in a way that is isolating and individual rather than collaborative.

Kathryn: While watching those videos, I found I started questioning student engagement and understanding when the class seemed very teacher or technology centered. I wonder, how much of this is true though and how much of this is personal bias based on what we value? I also remember finding it hard to watch videos of teachers that we had no relationship with because I think we were looking to understand what they were teaching, how they were teaching it and subconsciously we were trying to see if their values fell in line with ours.

Claudia: We were very conscious of content when watching the lessons. Our conversations sought to understand what the goal of the lesson was and what the students were expected to learn.

Kathryn: There were videos where I felt [as] if we had to struggle to figure out the goal whether or not the students knew what was expected of them. Do you think this transparency might be another value we hold as really important in our teaching?

Claudia: This is the value that we were most conscious of and the focus of our discussion. It was clearly what we were looking for in every video we watched and therefore our primary goal in determining what we considered an important factor.

Professional Growth in Terms of Practice and Identity

Claudia and Kathryn reflected on how their participation in this research project impacted their classroom teaching. Observing videos of themselves and other teachers gave them the feeling that the videos were opening the doors to their classrooms. Noticing elements of teaching from the videos permitted them to see common things they share with other teachers. It also took away their belief that they are the only expert of their classrooms. This later statement seemed to make them a believer of co-collaboration between teachers.

Claudia: This project allowed us to break down barriers and to feel safe in our teaching environment. This safety has allowed us to share our practices and create a learning community within our schools. Teachers often feel isolated in their own classes unwilling to share their strengths and weaknesses, in fear of being judged. Schools are set up to foster this practice since we can do our job behind closed doors with our students being the only source of feedback. That feedback is not even direct since the only measure is based on student success.

Kathryn: I think this is exactly it, this project really led me to understand that the richness of expertise lay with my colleagues and a new realization was that I had as much to share as they did. I have often found that as teachers we exist in silos of subject matter, grade or course and over the years we have been led to believe that unless we have those in common, we cannot collaborate. This project changes that for me.

Claudia: Teachers also have this antiquated concept of ownership of their lessons often unwilling to share. I think this comes more from fear of judgement. I remember thinking, "This lesson was designed for my students based on my style of teaching." But this thinking is limiting our students by giving them only one perspective. Math concepts should not be limited to only certain types of questions students are used to seeing. I find that students expect this! In fact, I have students who say to me, "We did not learn how to do this." In fact, they have, but they have never seen a question that resembled it so they assume that they do not have the conceptual knowledge to solve the problem. I love giving my students tests and quizzes written by other teachers since this tests their ability to apply their learning in situations they have not seen before. This new approach and clearer understanding came directly from experiencing the video project.

Kathryn: I was so guilty of this too and that fear of co-collaborating or co-creating came from my fear of judgement and a belief that nobody else could possibly be the expert on what went on in my classroom. Watching the videos, we did have me recognize that there were so many more similarities between my classroom (and students) and what was happening all across the country. It also made me realize that I had not been stretching my students far enough as I had not demonstrated what it meant to collaborate and build meaning. By having my students see me working with other teachers broke down some of their beliefs that they couldn't be the experts because they were just learning and also for my groups that they were just the "basic" Math group. I was becoming the expert AND, I was learning at the same time.

Kathryn: Coming back to this "You never taught us how to do this." comment is less of a comment about math skills but more that we need to continue to push our students to think. This is what changed for me in my classroom. I am not the expert or problem solver, I am the "host" of a thinking room where my students have to do the heaviest lifting. To do this I had to be confident enough to be comfortable with the silence and allow my lessons to exist partially in the unknown as they had to move with my students, not with a predetermined pacing.

The project helped the participants grow as teachers as it helped focus more on an important element in their classroom: student success. In fact, Claudia and Kathryn were able to generalize how they position students as learners. Instead of thinking that problems may be too difficult for students, they now tend to encourage students to solve them even though they might fail at first.

Claudia: I like what you said about the heavy lifting. This helps me feel more confident about pushing my students more. When my students complain that a problem was too difficult, I often second guess myself and think that maybe they're right. Instead, I should be encouraging them, and teaching them strategies they need to help them break down the concepts rather than "dumbing it down" to use problems that resemble examples from their workbook and just change it up a little. We all want our students to be successful, but how do we define success? It's up to us to redefine success in our classrooms and we need to incorporate more formative evaluation situations to achieve this.

Kathryn: I remember a colleague once saying that she never identifies problems as being easy or hard and for a long time, I thought she was crazy, how could she not warn the students in advance of how much work or difficulty a problem would be. Now I get it, all the problems should be presented as achievable and possible, it might just mean a new approach, new tools or collaboration might be necessary. I have been working hard to refine success as [a] problem solving process and not just [an] end result.

Claudia: I never thought of it that way! Even now I'm learning something new. I really like this since just this week I gave my students a very challenging question and I told them it was difficult followed by, "But you are all capable of solving it." I guess I should have skipped the first part since they all looked at me discouraged before they even began except for those students who like a good challenge. Unfortunately, they are not the ones that need to be motivated. There is no point in telling them that the problem is difficult. I think we tell them this because we feel we need to prepare them for failure and in fact what we are actually doing is preparing ourselves for failure.

Claudia: I think this risk taking that you are working to develop in your class very much mirrors the change in our teaching as a result of this project. We are more willing to try something new, exist in the less comfortable places, talk through our thinking and reach out to others when we're not sure where to go next. I think the more we go on this journey the more our classrooms evolve and mirror this too.

Kathryn: As teachers, we are often our own harshest critics but rarely ever have the ability to view ourselves from our students' perspective. Reflection has always been a regular part of my teaching routine happening before, during and after lessons. Being part of the video project represented a shift in my reflection as I was able to reflect while viewing my lesson. This provided me the ability to reflect from an outsider's perspective and I was able to see how I was truly interacting and connecting with my students not how I believed I was. This "flipped" perspective really challenged my beliefs of myself as a teacher. At first, I was quite critical of my phrasing, questioning practices, and affirmations I gave to my students. The longer I watched though, the more I was able to distance myself from self and take in the video for what it was showing and not the little things about myself I was nitpicking about. I began to see my class as a being in and of itself and not with the teacher as the main focus. This more holistic perspective allowed me to see that my students were engaged and participating in the lesson and the values of safety and inclusion that I strive to cultivate in my class were ever present. I became aware of little routines in my classroom that I take for granted as they were established so long ago, and how effective they are in creating independence for my students.

Claudia: Watching videos of myself with my colleagues was an empowering experience but also required a lot of vulnerability. I believe my view of my role as a teacher and values associated with this role are constantly evolving so to let someone see you in a form of identity that is still being built is very intimidating and scary. Having people watch videos of themselves in this same vulnerable state allowed for us to give true feedback drawing on the pre-existing relationships and friendships we had. Our relationships created a safe space for this vulnerability and then permitted a very open dialogue of questioning and commenting. Through these interactions, I believe I truly grew as a teacher.

Kathryn: One thing that changed in my classroom is that as a teacher I now feel comfortable enabling the meaningful struggle of my students. It's okay to not get the answer right away and it is productive to have them do the heavy lifting. This is true regardless of the ability-level, age or grade of my students. The students I am currently teaching struggle with a range of learning disabilities and challenges which is why they are attending the school I am working at. The school is quite small, and my classes are organized based on curricular competency not grade or age. In my classroom this year, I have a portable whiteboard that sits in [a] corner of the room. My students have named it the "pause board." We use it when we decide to "pause" a lesson. This pausing used to be called by me to investigate an interesting question or dig deeper into content, but it is always spontaneous and guided by the responses my students are giving me. Just the other day though, one of my students said, "let's pause this and go to the pause board," because at 11 years old, she recognized that the question she asked was one that extended the learning of the whole class. Not only do I feel that I have been able to be more spontaneous to move with my students' needs but also comfortable to let them be active seekers of knowledge.

Kathryn: And added authenticity to teaching by keeping lessons open-ended and less scripted and encouraging student input. This sets up a climate for authentic learning. Students need to feel safe to make mistakes. How do we create a safe learning environment unless teachers are willing to take risks themselves in front of their students?

Professional Growth: Shift from Teacher-Participant to Teacher-Researcher

Although the participants do not seem to be at ease with the term *teacher-researcher*, the project made them shift from a teacher who is curious and looks into finding solutions to problems instead of finding excuses or people or things to blame. These teachers contrasted the project sessions with teacher meetings at school, which they find unproductive because of the lack of solution finding. Claudia clearly stated that she now approaches teacher meetings at school in a different way, by having clear goals in mind and directing discussions toward finding solutions related to these goals. However, the participants also recognized that this process of finding solutions cannot be done alone. They perceived that this must be a collaboration between fellow teachers and even the school administration.

Claudia: I think my willingness to take risks in the classroom also extended beyond the walls of my classroom as I found myself seeking out more opportunities to take on new initiatives and deepen my understanding of evaluation and curriculum. I still feel slightly uncomfortable with the title of teacher-researcher as my definition of the term is still under construction. My belief about being a researcher was always associated with a highly educated individual conducting experiments in a more scientific manner. I think the curiosity of researchers is the piece that I feel fits my professional growth. I feel now that I am an active observer and questioner of practices, my own always in the forefront.

Kathryn: I noticed that once our video group had established trust after we had been vulnerable by watching ourselves and others, our conversations with our colleagues shifted from being gripey student-centered topics to more solution driven topics. For example, I remember being guilty of complaining about Johnny not paying attention or doing his homework but after this project the conversations shifted to how did you teach this to keep it interesting or do you assign homework every night?

Claudia: In the past, department meetings often focused more on problems and less on solutions. We share the struggles we have with the students and often find excuses for our failures. The lack of motivation is placed on the students alone and the gaps in their learning is [sic] often blamed on the previous years. "What are they teaching them in elementary school? How is it that they still don't know how to....?" I find myself having less tolerance for these kinds of exchanges. What has changed is how I approach my department meetings. My goal now, as a coordinator, is to shift the focus to constructive, solution driven dialogue. I make sure I have clear goals in mind where the results should include something the teacher can take away with them (sharing lessons and practices that work). Trust is built in a team when there is a realization that we need each other to be successful and that none of us can achieve success alone. Sharing our weaknesses and asking for help is empowering. Is this not what we try to encourage our students to do? I now refuse to entertain any whining as it is unproductive and time consuming. I quickly try to shift the discussion to focus on our goals. It is very rewarding for all participants to end a meeting with something they can use in the classroom.

Kathryn: I absolutely agree but I find colleagues are a little reluctant to let go of the excuses and move to an entirely solution-oriented approach because it truly required everyone in the process to be vulnerable and take a hard look at themselves. I think we are more comfortable in this role because of our work with the videos we watched. The biggest challenge I am facing is trying to recruit teachers (in my new school) who want to collaborate. The feedback I keep getting is there isn't enough time. I am persistent though as I know this collaboration is something I need in my growth as an educator.

Claudia: There is no question that it is challenging to find colleagues willing to move to an entirely solution-based approach. For every success story I have, I also have those that have gone sideways. Collaboration for some means something completely different. There are those teachers who are very vocal and others who have little to say. It is important for a group leader to create an environment where everyone at the table has a voice. This for me is the greatest challenge. The video project probably unintentionally gave each of us a voice.

Kathryn: I think that's the other interesting thing, we are also comfortable not knowing the answer and understanding it may be a process (or series of failed attempts) before we get there. This requires the same level of trust and vulnerability because team members have to be able to critique the solution and process without doing the same to each other. This process really has had me tap into looking at the group dynamics, social psychology and needs of each group member more than ever before. I think being able to do this has to do with a clear[er] understanding of who I feel I am as an educator.

Kathryn: Being in a leadership role can also be different things for different people. Teachers, in particular, are used to being in a leadership position with their classes and find it difficult to take a back seat when working with other adults. The video project forced us into a vulnerable position before we even met as a group.

Claudia: The video project taught me that a good leader requires the ability to get past the noise and read between the lines making sure everyone's voice matters. The researchers leading this project modeled these leadership skills by creating an environment where everyone's professional experience mattered and that we were all there to build our understanding.

Kathryn: This is so important; I really enjoyed the process of the video project because I always felt my voice mattered. I think this is the biggest take home from this project and permeates in every aspect of my teaching career.

Kathryn: We were/are learners before teachers. Our curiosity and love of learning is what drives us professionally: new initiatives, conferences, writing exams, department coordination.

CONCLUSION

Although participating in the research project was a risk for these teachers, their written comments highlighted that the experience enabled them to open the doors of their classrooms through video by sharing their classroom practices. In doing so, they were able to observe and notice pedagogies of teaching mathematics. Feeling vulnerable at first, the two teachers, Claudia and Kathryn, were able to realize that they shared common elements with their colleagues and teachers from other regions. This realization appeared to be surprising for Claudia and Kathryn as they felt they were alone in facing the challenges in their own classrooms.

As for the Learning to Notice framework, the post-research conversation of these two teachers showed the first and third elements of noticing (identifying what is important in a teaching situation, and making connections between specific events and broader principles of teaching and learning). They were able to highlight elements from the video-recorded lessons that were important to them, such as their planning and success in their classrooms, and extrapolate with reflections that revealed changes in their values as teachers. A powerful reflection was their talking about planning lessons. For instance, Kathryn realized that a crucial aspect of building a good lesson involves outlining good questions to ask students. To Kathryn and Claudia, their reflections made them realize that it was the quality of the questions they posed to students that improved the quality of instruction in mathematics. As for success in the classroom, these teachers, reflecting after observing lessons in which teachers challenged students with difficult tasks, realized the importance of not judging tasks as simple, appropriate for their students, or too difficult, but rather, positioning students competently as learners by encouraging them in their process. Conversations about students' reasoning and learning, the second element of the Learning to Notice framework (using what one knows about the context to reason about a situation), were absent in the teachers' conversations. This is probably based on the goals of the research project itself, which focused on the pedagogies used by teachers and not necessarily on student reasoning.

This research project seemed to enhance the professional growth of these teachers. Although Claudia and Kathryn do not seem totally comfortable with the label teacher-researcher, they felt they grew into this role by developing certain elements of research, including curiosity and the desire to find solutions. They even seemed to be able to bring this teacher-researcher perspective into other professional contexts, such as department and staff meetings. Their discovery to find solutions seemed to Claudia and Kathryn more productive than putting the blame on someone or some things.

Throughout the project, informal feedback to the research team members suggested that teachers saw their own professional growth through participation. Kathryn and Claudia's interest in writing reflections on their participation in the project provided us with an opportunity to think more about this dimension of the project. Based on their written conversation, the Observing Teachers research project seemed to bring many opportunities for teacher professional development. It is not possible to make strong conclusions based on the small number of participants, but more extensive research should be conducted to further explore this alternative as a professional development opportunity.

REFERENCES

Ball, D. L., & Cohen, D. K. (1999). Developing practice, developing practitioners: Toward a practice-based theory of professional education. In G. Sykes & L. Darling-Hammond (Eds.), *Teaching as the learning profession: Handbook of policy and practice* (pp. 3–32). San Francisco, CA: Jossey Bass.

Brown, J. S., Collins, A., & Duguid, P. (1989). Situated cognition and the culture of learning. *Educational Researcher, 18*, 32–42.

Chi, M. T. H., Glaser, R., & Farr, M. (1988). *The nature of expertise*. Hillsdale, NJ: Erlbaum.

Frederiksen, J. R. (1992). *Learning to "see:" Scoring video portfolios or "beyond the hunter-gatherer in performance assessment."* Paper presented at the annual meeting of the American Educational Research Association, San Francisco, CA.

Glaser, R., & Chi, M. T. H. (1988). Overview. In M. T. H. Chi, R. Glaser, & M. J. Farr (Eds.), *The nature of expertise* (pp. xv–xxvii). Hillsdale, NJ: Erlbaum.

Goodwin, C. (1994). Professional vision. *American Anthropologist, 96*, 606–633.

Goos, M. (2008). Sociocultural perspectives on learning to teach mathematics. In B. Jaworski & T. Wood (Eds.), *International handbook of mathematics teacher educa-*

tion: Vol. 4 The mathematics educator as a developing professional (pp. 75–91). Rotterdam, the Netherlands: Sense Publishers.

Larkin, J. H., & Simon, H. A. (1987). Why a diagram is (sometimes) worth ten thousand words. *Cognitive Science, 11*, 65–69.

Lesgold, A., Rubinson, H., Feltovitch, P., Glaser, R., Klopfer, D., & Wang, Y. (1988). Expertise in a complex skill: Diagnosing X-ray pictures. In M. T. H. Chi, R. Glaser, & M. Farr (Eds.), *The nature of expertise* (pp. 311–342). Hillsdale, NJ: Erlbaum.

Schoenfeld, A. H. (1998). Toward a theory of teaching-in context. *Issues in Education, 4*(1), 1–94.

Sherin, M. G. (2001). Developing a professional vision of classroom events. In T. Wood, B. S. Nelson, & J. Warfield (Eds.), *Beyond classical pedagogy: Teaching elementary school mathematics* (pp. 75–93). Hillsdale, NJ: Erlbaum.

Shulman, L. (1996). Just in case: Reflections on learning from experience. In J. A. Colbert, P. Desberg, & K. Trimble (Eds.), *The case for reflection: Contemporary approaches for using case methods* (pp. 197–217). Boston, MA: Allyn & Bacon.

van Es, E. A., & Sherin, M. G. (2002). Learning to notice: Scaffolding new teachers' interpretations of classroom interactions. *Journal of Technology and Teacher Education, 10*(4), 571–596.

van Es, E. A., & Sherin, M. G. (2008). Mathematics teachers' "learning to notice" in the context of a video club. *Teaching and Teacher Education, 24*, 244–276.

CHAPTER 10

RESEARCHING MATHEMATICS PEDAGOGY

Richard Barwell, Lisa Lunney Borden, Dominic Manuel, David A. Reid, Annie Savard, Elaine Simmt, and Christine Suurtamm

The Observing Teachers project involved working with groups of middle school mathematics teachers in different regions of Canada to construct videos representing their practice. Groups then worked together to select videos that were typical of their region to be sent to teachers in other regions. The same groups then discussed the videos received from other regions. This research design drew on enactivist principles. In this chapter, we draw on these principles to reflect on some of the challenges that arose in implementing the project. In particular, we refer to three aspects of enactivist research highlighted by Simmt and Kieren (2015): the role of observers, the relationship between ontology and epistemology, and the role of ethics. Thinking about these ideas led us to notice four emergent themes relating to the process of enacting the project: researcher-teacher-learner relationships; unexpected events; learning to be a researcher; and insiders-outsiders. Underlying these themes is the importance of relationships in the conduct of research with mathematics teachers.

INTRODUCTION

In this chapter, we reflect on and theorize some of the many challenges that arose during our work on the Observing Teachers project. The project was particularly complex as it involved multiple researchers and multiple participants in different

Researching Pedagogy and Practice with Canadian Mathematics Teachers, pages 213–230.

regions of Canada, each with their own particular points of view on the nature of mathematics, teaching, learning, and research. Indeed, the overall conceptualization of the project was designed to permit us to describe the nature of mathematics pedagogies in different regions of Canada, while recognizing that each of us came to the project with implicit assumptions about this topic. By inviting teachers to observe teaching from regions of Canada other than their own, using videos of lessons selected by teachers in these other regions as, in some way, representative of teaching in their own region, we aimed to examine how these observers would reveal something of their own assumptions about teaching mathematics. This same logic, of course, applies to us as members of the research team. That is, our own observations of the various groups of teachers discussing videos from their own regions and then from other regions are necessarily revealing of our own assumptions about mathematics pedagogy and about research (among other things). Our assumptions are revealed in our implementation of the research design, our interpretations of the data, and our interactions with participants and with each other. In this chapter, then, we examine some of the challenges and dilemmas that arose as a result of our reflexive research design. By making these challenges and dilemmas explicit, our goal is to offer points of reflection for researchers embarking on similar forms of research in mathematics education. We believe that such points of reflection can be productive in the development of our practice as researchers. Reflecting on the issues we discuss in this chapter was certainly productive for us in the development of our own practice.

We begin with an oversimplified history of anthropology. Early research in anthropology involved outsiders observing and describing "exotic" cultures, often with a critical and paternalistic eye. Whether non-European cultures were portrayed positively, as "noble savages," or negatively, as "primitives," they were definitely viewed as "other." As anthropologists became aware that their observations were strongly colored by their own cultures, they began to focus on measuring material artefacts and other aspects of culture that could be turned into "objective" numerical data, meeting the hoped-for standard of a science. The goal was to bracket out the outside observer's own biases, by reporting only data thought to be immune from influence by those biases. This approach too was critiqued, as it meant discarding many of the essential, but not measurable, aspects of culture. In parallel, participant observation emerged as an approach to studying other cultures by participating in cultural practices, attempting to see them as both an insider and an outsider.

Research in mathematics education has followed a similar path. The first academics involved in mathematics education were not classroom teachers, but were largely professional mathematicians or psychologists who analyzed curriculum and teaching practices from outsider perspectives[1]. A desire for more objective

[1] See, for example, this "Historical Sketch of ICMI": https://www.mathunion.org/icmi/organization/historical-sketch-icmi

methods led some to concentrate on quantitative techniques, but as in anthropology, this was criticized as leaving out essential aspects of learning that could not be measured. Some mathematics educators turned to participant observation, either as teacher-researchers observing in their own classrooms, or as researchers working in cooperation with teachers. This shift has led to new epistemological perspectives, as researchers have grappled with questions about what it means to know another culture or to make claims about different ways of teaching.

RESEARCHING AS A REFLEXIVE PROCESS

The overarching design for this research project drew on enactivist principles. Enactivism has become influential in mathematics education research, particularly in Canada (see Reid, Brown, Coles, & Lozano, 2015). Although enactivist principles encompass theory, methodology, and epistemology, for this chapter, we focus particularly on the methodological dimensions as a way to make sense of some of the challenges that arose during the conduct of the overall research project. We draw not only on formally collected data, but also on the less formal interactions inherent in any research project. These less-formal interactions include our discussions with participants, the interactions within our regional teams, and our regular meetings as a national team. Much has been written about enactivist approaches to mathematics education research (see Reid et al., 2015, for an overview). For this chapter, we draw on three specific aspects or methodological "moves" highlighted by Simmt and Kieren (2015), namely the role of observing and the observer, the nature of meaning-making as bringing forth a world of significance, and ethics.

Observing and Observers

A key guiding tenet for the study presented in this book is Maturana's (1987) dictum that "everything said is said by an observer." What this means is that it is not possible to make claims about the world that are somehow separate from the person making the claim—that is, the observer. This principle has fundamental implications for research about learning or teaching mathematics (or anything else). Research is often thought of as a process of making claims about learning or teaching based on careful analysis of data. The inherent reflexive connection between the observer and the observation, between the knower and the known, recasts this process. Given this perspective, in the Observing Teachers project we were faced with the challenge of researching mathematics pedagogy when we, as researchers, do not have direct access to what mathematics teachers perceive as characteristic of that pedagogy. There is a solution, however, provided the relationship between what is said and the observer who says it is clearly understood. As Simmt and Kieren (2015), referring to work by Maturana (1987), point out: "at the heart of any useful 'scientific' observation is a means … by which the listener to the observation is enabled to understand how the observer 'sees' the situation" (p. 308). That is, what people say tells us something about the person saying it. In the Observing Teachers project, we applied this idea in the design of the project, by soliciting teachers' observations of each other's teaching, in order to understand something about how the participating teachers "saw" mathematics pedagogy. Of course, this principle also applies to us; we (the researchers) are observers of the observers (the participating teachers). This means that you, the reader, can learn something about how we, the researchers, made sense of the ways that the participating teachers made sense of mathematics pedagogy. At times, these layers of observing, although elegantly conceived within the research design for the project, led to some particular challenges.

Bringing Forth a World of Significance

From an enactivist perspective, actions are determined by the structure of the actor in interaction with the environment. For example, at a given moment in a mathematics class, a teacher acts in a way that is determined by their structure; i.e., the historically accumulated ways of interpreting what students do. This structure changes over time as a result of experience. The reflexive relationship between acting in a given moment, and the structure of the person acting, gives rise to patterns of interaction between individuals or contexts, which are known as structural coupling. For example, the behaviors of students and a teacher in a mathematics classroom adjust to be in some sense mutually aligned, although both will change over time.

Of particular importance is the idea that knowledge is not a fixed entity inside someone's head; rather, *knowing* is contingent and situated within a context. Indeed, knowing is jointly produced through a process of interaction that involves the use of language or other signs. This process of "jointly doing knowing" through interaction results not in the description of the world (a regurgitation of knowledge), but rather in the bringing forth of the world in each moment. "Jointly doing knowing" means the collective process of enacting knowing through interaction. Thus, in the Observing Teachers project, researchers sat with participating teachers to discuss videos of other teachers teaching mathematics. In these interactions, teachers were not providing objective descriptions of the videos, and researchers were not making objective observations of the teachers. Rather, the group is, in the moment of the focus group, bringing forth a world of meaning relating to the videos. If, for example, the group agrees that some particular pattern of teaching is a common feature in the videos from teachers in a particular province, we see this as a process of jointly producing this feature from the infinitely many details of the videos and the participants' interactions. This process is, of course, conducted through language, which, among other things, serves as a way to clarify and coordinate each individual participant's noticings. Again, there is a layered aspect to the study: as researchers, we also worked together (in subgroups or as a team) to make sense of data collected from the regional groups of teachers. One important implication of this perspective is that our researcher interactions involved us in bringing forth a world of significance in relation to mathematics teachers bringing forth a world of significance about mathematics teaching in their collective discussions about the videos from different regions.

Ethics

The ethical dimension of research, and of human relations in general, arises from the previous set of ideas about the structurally coupled processes of bringing forth a world. To explain the idea of bringing forth a world, Simmt and Kieren (2015) cite Maturana (1988): "Everything that we do becomes part of the world that we live in as we bring it forth as social entities in language. Human responsibility ... is total" (p. 40). By jointly bringing forth a world, we share a responsibility for everything in that world. Varela (1999) devoted a series of lectures to this topic. In them, he described "immediate coping," in which "the situation brought forth actions from us" (p. 5). These actions stem from the history of our structural coupling with others, and with an environment, and are habitual, in the sense that we are often not aware of making choices or decisions, but are simply being (doing, knowing) in the situation. In terms of ethics, if we see someone fall, we may, in our immediate coping, go to help that person up. We have been socialized into a human world of ethical significance through which such behaviors may become "immediate." Varela contrasts immediate coping with "deliberate, willed action" (p. 5), in which we make "conscious" decisions after due reflection. For example, we might contact municipal authorities to report a dangerous part of the sidewalk so that people do not fall in the future. For Varela, deliberate, willed action, in which we analyze a situation before acting, is largely used in unfamiliar situations—situations for which we have not yet developed a repertoire of immediate coping behaviors. Neither form of interaction is fully ethical. Immediate coping is largely habitual and hence unreflective, while deliberateness often leads to an over-rigid application of rules or principles. Both can be ineffective in complex situations. Varela argued that a middle position is the basis of "ethical know-how." Brown and Coles (2012) refer to this position as "non-deliberate readiness-in-action," which they describe as a form of immediate coping with awareness of the motivations that prompt our actions. This approach requires "an on-going alertness to the detail of what we experience" (p. 223). This non-deliberate readiness-in-action is the basis of ethical know-how:

> Truly ethical behavior does not arise from mere habit or from obedience to patterns or rules. Truly expert people act from extended inclinations, not from precepts, and thus transcend the limitations inherent in a repertoire of purely habitual responses. (Varela, 1999, p. 31)

In the Observing Teachers project, we experienced all three modes of interaction: immediate coping, non-deliberate readiness-in-action; and deliberate, willed action. In some situations, it is clear in retrospect that habitual immediate coping kicked in. In the course of conducting the group discussions, for example, unexpected directions sometimes led research team members to fall back on their repertoire of interviewing behaviors. At other times, we engaged in extended collective reflection about the research and our experiences of implementing the project, particularly during project meetings. We thought and worked deliberately to develop shared interview protocols, for example. On some occasions, non-deliberate readiness-in-action could be inferred, as when awarenesses arising from our project discussions became present in the moment of interaction with research participants, leading to smooth decision-making consistent with project principles, but responsive to the current interaction.

In the rest of this chapter, we share some of the methodological issues that became significant in our conduct of the study. We focus, in particular, on four themes, drawing in each case on examples from the research project presented in the form of first-person vignettes. The common underlying theme is our concern for relationships throughout the project.

RESEARCHER-TEACHER-LEARNER RELATIONSHIPS

As in any complex process, we were each aware of occupying multiple roles during the course of the project. Even within the team, at different times, members led discussions, proposed analyses, or were participants in discussions about the project. During data generating activities involving participants, these roles expanded to connect our research identities with our other professional identities, particularly as teachers. Sometimes these identities were in harmony, in other situations a tension was felt. The first reflection speaks to the importance of the intermingling of these roles in our work on the project.

Christine on Bringing Practice into Research

Much of my work includes training beginning researchers; often when they are asked to comment on ethical issues, they restrict themselves to speak about the questions and responses that are prompted by a research ethics board application form. It is only after doing research for some time that one realizes that every moment of doing research is an ethical moment. The way a researcher recruits participants, the approach the researcher takes on first meeting the participant, the facial expressions of the researcher as he or she stands behind the video camera capturing the teacher's interactions with students, and the way the researcher interacts with the teacher as the lesson or interview finishes, all portray an ethical stance and determine whether the participant feels comfort or judgment.

Engaging in qualitative research is about engaging in relationships, which brings forth values such as trust, ethics, integrity, and respect. The Observing Teachers project leans on research relationships between and among researchers, teachers, and graduate student research assistants. Building relationships between researcher(s) and teachers is at the heart of the research that I do. I value the work of teaching. Teaching is part of my own work and it is not easy work. When I work with teachers, it is important that I build a trusting relationship and that the teacher knows how valuable and complex I think their work is. I am not there to judge but to better understand those moment-to-moment decisions that teachers make as they build relationships with students and interact with mathematical ideas.

So often we hear proclamations of bringing research to practice. My view in doing research is focused on the reciprocal relationship between research and practice. It is not so much about bringing research to practice as it is about bringing practice to research. Or rather, the research seeks to understand the complexity of teaching, to examine those reflective moments and actions that teachers make, and to consider why they make them. It is about respect for the work that teachers do.

I often lean on the idea of both teacher and researcher having a stance of inquiry (Cochran-Smith & Lytle, 2009). Both are trying to gain a deeper understanding of teaching. For the teacher, it is inquiry into their own teaching—it is often those thoughts and replays that the teacher engages in as he or she drives home from school. Those thoughts, such as "When Josh asked that question, I'm not sure I gave him the answer that helped to move his thinking," or "I wonder if Sarah felt defeated when I suggested she try a different strategy," plague teachers. However, they also provide them with the opportunity to revisit the day and to reflect on what they did and why, and to think about the instructional moves they will make the next day.

As a researcher, I also adopt a stance of inquiry and do so on many different levels. As I conduct research, my research questions guide my inquiry, but further than that, I am constantly inquiring about the act of doing the research itself. I revisit the ways that I asked interview questions, responded to the interviewee, captured particular classroom moments on video, and conducted myself as a researcher.

Comment

Christine's reflection on her participation as a researcher in the Observing Teachers project reflects well the ideas and orientations summarized at the start of the chapter. In particular, her stance involves both teachers and researchers engaging together in inquiry. This stance is in contrast to an approach in which researchers seek to describe and analyze teachers' practices or experiences. In Christine's reflection, the mutual observation of each other and of the classroom is part of jointly bringing forth an understanding of what is going on in the videos. Both teachers and researchers contribute to this process in a way that cannot be reduced to separate observations, distinctions, or understandings. Rather, understanding is a joint outcome of the interactions between all participants in relation to the videos. This means that these understandings reflect, in part, Christine's distinctions and experience of mathematics teaching, mathematics classrooms, and so on. They reflect the teachers' distinctions and experiences of mathematics teaching and mathematics classrooms. Christine's stance also reflects the ethical position of non-deliberate readiness-in-action. Through constant inquiry into the research process, including discussions with the research team, Christine illustrates how researchers can avoid "mere habit" and can refine the sensibilities in order to act with sensitivity in working with teacher participants.

UNEXPECTED EVENTS

All research is contingent to some extent, as researchers adjust and adapt to circumstances. How we do this is revealing of how we perceive things. Throughout the project, during research team meetings, unexpected events were the focus of discussion. Participating in these discussions often led us to realize that our perceptions of the research project were sometimes quite different. The following vignette describes an unusual and frightening event that interrupted data collection in New Brunswick, Canada.

Annie and Dominic in Lockdown

We had planned to go to New Brunswick to conduct focus groups within a period of four consecutive days. During the first three days, teachers would observe and select the videos to be shown to teachers in the other provinces. On the last day, they would observe videos from other provinces. The focus groups were conducted on a university campus. Two of the teachers lived close to the university, while two others travelled from elsewhere in the province and stayed in university residence accommodation, as did we.

On the evening of the second day, there was a shooting in town: a man shot three policemen. Because of the shooting, all schools, most businesses, and the university campus were closed, and a big part of the city was on lockdown throughout the third day. The next morning, Annie wanted to go to a restaurant because the cafeteria on campus where we had breakfast every morning was closed. The restaurant was, however, in the lockdown zone. Annie thought of walking to a McDonald's restaurant instead, but would have needed to walk through a forest area where the shooter was hiding. Dominic wanted to ask one of the teachers to fetch us, but Annie pointed out that it was not ethical to ask a teacher to put his/her safety at risk. We needed to get out, but our safety was a concern. Another participant offered to go to a store to get us some food, but we declined because we were concerned for her safety. In the end, one of the participating teachers staying in the residence decided to go for breakfast and offered to take us all. That solved our first ethical dilemma: the offer came from the teacher; it was not requested from us.

During breakfast, we looked for solutions in order to continue the focus group, given the fact that we were losing a day of data collection, and one of the participating teachers had to leave at the end of the day. This was a difficult situation, because the videos needed to be selected before we returned to Montreal. We opted to split the third day of research, observing the exemplary lessons, into two days. The first day would be done with the two teachers staying at the residence, and the following day would be with the two local teachers. In this way, the teacher who needed to leave would be able to participate in the whole first phase of the research. We also postponed the second phase of the project for this region until the fall[2].

We had to conduct the focus group in one of the dormitory rooms because the campus buildings were closed. At the end of the process, the two teachers selected the videos they wanted to show to teachers in the other provinces. The next day, we agreed not to tell the remaining two teachers which videos the first two had selected so as not to influence their decision. The school district had announced the day before that schools would be closed on this day as they had not caught the shooter yet. This was a challenge for one of the participants because she was alone with her child and the daycare was closed. She was able to get a babysitter, but only for four hours, so we had to conduct this focus group at a faster rate.

At the end of the final day, both groups had agreed on the same videos to represent an introductory lesson on fractions and an exemplary lesson. Opinions were divided between two videos for a typical lesson. We thus needed to have another short meeting so the participants could make a decision about the typical lesson video. Because Annie and Dominic had their flights reserved at the end of the fourth day, and because the focus groups were held at the end of the school year, we had a short Skype meeting with the four teachers the following September to agree on one of the lessons.

Comment

In this alarming situation, Annie and Dominic were clearly (and understandably) in immediate coping mode. They had little previous experience or reflection on which to draw in their decision-making as the situation was so unusual. They did, however, keep in mind ethical principles, such as their concern for participant safety, which informed their thinking as the day unfolded. Their account also gives a sense of how, in a fluid and uncertain situation, the nature of the situation emerged through their interaction with each other, with other participants, and with news and social media. We get a sense of how their understanding of what is sensible and what is best avoided were in flux in response to new information. The decisions they made as the day unfolded were highly contingent. It is interesting, for example, that they were determined to complete as much of the data collection process as possible, despite the circumstances, and that their subsequent actions can be understood with respect to this goal. It is also noticeable that they had different individual awarenesses, based on their different histories. Dominic knew the region and the participants well whereas Annie did not. So, for example, when

[2] The first phase of the project included teachers viewing videos of their own teaching; the second phase included teachers viewing videos of teachers from other provinces.

Annie wanted to walk to get breakfast, Dominic was concerned that she would pass through the area where the shooter may be hiding.

LEARNING TO BE A RESEARCHER

There was a range of expertise in the research team, including seasoned professors and doctoral student researchers occupying different roles. Dominic, originally from New Brunswick and studying in Montreal with Annie, was a doctoral researcher who became an important member of the team. His contributions included data collection in more than one region, as well as recruiting teacher participants in New Brunswick. Subsequent to the end of the data collection, he completed his doctorate and obtained an academic position. His journey through the project, therefore, offers some insight into a transition from "legitimate peripheral" researcher to greater autonomy and expertise.

Dominic's Journey

During my graduate studies, I participated in many research projects as a research assistant. The Observing Teachers project was very beneficial for me, not only because I used some of the data for my doctoral research, but also because of the various opportunities it brought, such as working with multiple researchers from across Canada, collecting and analyzing data with groups of teachers from different regions, and using focus groups as a method.

When it comes to doing research with teachers, I first realized that the project involved more than simply collecting, interpreting, and analyzing data to enable the advancement of knowledge in education. When participants from the three groups of teachers from which I collected data mentioned that this was the best professional development they received in their profession, I realized that educational research should be more than merely data collection, analysis, and dissemination. I now see doing research with teachers as a collaborative process where both the researchers and participants work together as a community, and all its members should meet their needs and goals. In other words, it is doing research *with* and not doing research *on* teachers. Of course, the researchers' main goal is to collect various types of data to inform the researchers' community. However, I argue that teachers must get something out of the process for their professional growth. Researchers and teachers should all work together and share their expertise. Researchers are experts in various theories and fields in education and in conducting research, but teachers are experts of their classrooms (see, for example, Jaworski, 2003).

Secondly, to build this collaborative research community, I value the importance of creating a safe environment for all participants. This involves not only creating a space in which all participants can freely share their values, reflections, and opinions about teaching for data collection, but where the researcher also uses proper judgment to assure the safety of the participants. I recruited a group of teachers for one region. I knew those teachers very well and felt like an insider when I was collecting data with them. During the first day of focus groups, I was alone with the teachers and felt their safety in sharing their views on teaching. During the second day, however, Annie joined the group. Annie was an outsider as she did not know the participants, and the dynamic changed. Efforts had to be made to rebuild the safe space with the researchers. For phases of the project conducted in other regions, I was an outsider, as I did not know the teachers. I even joined one group only during the second phase. This insider or outsider relationship with the participants made me question the validity of the data, especially in the context of focus groups. What effect does knowing or not knowing the other participants have on the data that participants are willing to share?

According to Berliner (2002), educational research is the hardest science of all, due to many factors, including the difficulty of generalizing results based on classroom settings. From my reflection, added to the challenges of using methodologies that will bring accurate results, I argue that researchers should create safe collaborative communities with the participants, assuring the members' security and freedom to share their ideas. Creating such conditions may bring equitable growth for all members of those communities. This may not be possible in every context, but I suggest that these two ideas be implemented as much as possible. Now that I am at the end of my doctoral studies and have been appointed to an assistant professor position, I will bring those lessons with me into my profession.

Comment

In Dominic's account, he describes his changing perceptions about what researchers (can) do and what their role can be when conducting research with classroom teachers. It is interesting to compare his summary of his initial perceptions with those of Christine earlier in this chapter, who is a researcher with long experience of working *with* teachers. Dominic makes a new distinction (in his world) between research *with* and research *on*. This distinction emerges from observing and interacting with other members of the research team and leads to a new way of being and doing in the context of research with classroom teachers. This account is an illustration of how reflection on initially habitual responses, and on behavior largely guided by our established research protocols, leads to new ways of being and doing, particularly when shared with other members of the team. Dominic does not simply come to see doing this kind of research differently; it becomes different—his world has changed. This process is continuous, of course, with Dominic identifying goals for his future work, which will lead to further opportunities for reflection and new forms of research practice.

INSIDERS-OUTSIDERS

Our final theme concerns a long-standing issue in qualitative and ethnographic research, particularly when that research is focused on culture(s), including regional forms of mathematics pedagogy. This issue concerns the nature of a researcher's access to a culture if they are or are not members of the community that reproduces it, often referred to as the insider-outsider question. As illustrated in the two reflections that follow, the insider-outsider dichotomy is simplistic: researchers may be a bit of both, and the question of who constitutes an insider is also complex. A researcher with teaching experience may be considered an insider to the teaching profession, but an outsider to a particular school, for example. The question of who counts is a version of the observer principle: Insiders are insiders if people who share a culture perceive them as such. Questions also arise about the value and legitimacy of research conducted by researchers perceived as outsiders.

Elaine on Moving between University and School Settings

As I write this, I am trying to understand why I interacted with the teachers in my region, Alberta, the way I did. Although we, as a research team, had agreed on a process, I found myself working outside of our agreed upon processes. Recall that we video recorded three lessons, then asked teachers to watch their videos and select 20-minute clips to share with teachers from other regions in Canada. They were then asked to take those clips and watch them with the other members of the research study to choose three lessons from among the collection to share with others. At that selection process, we were to video record the teachers discussing the selections and use that as data for our study. Then they would watch the videos of teachers from other regions and that discussion would also be recorded to be used as data for the study.

None of this went as planned. The first violation/mutation of the process involved a research assistant taking the hour-long lessons and extracting 20 minutes from the tapes that focused on the teachers' actions. Then those tapes were offered to the teachers to individually select from to share in their regional group. The second was the teachers deciding they would like to create a mix that involved multiple lessons. The third was that the facilitator (I just wrote that word rather than *researcher*—what does that say) was an active participant in the discussions, posing questions, commenting, and sometimes explaining. I offer a brief account of my history as a teacher, teacher educator, and researcher to help unpack some of my actions.

My perspective within the regional pedagogies project was based on my history of interactions in schools. I am a former secondary school teacher and researcher, and I have conducted multiple projects in mathematics classrooms with local teachers in the region I studied. Some colleagues in the project view me as an insider in relation to the classrooms and teachers who participated in my region. Their assertion is likely accurate. I have participated in curriculum development in this province, I am a teacher educator in the largest pre-service program in Alberta, and I have been involved in hundreds of hours of professional development for teachers in this region. I am up-to-date on teacher quality standards and certification requirements, and I understand the distinct roles of the Ministry of Education and the professional organization that impact the day-to-day lives and instructional practices of mathematics teachers. Although my own teaching and teacher education experience is firmly rooted in this region, for the last fifteen years, I have been involved in international education projects; most of this work has been in East Africa, South Africa, and the Persian Gulf.

My insider status is more than just the fact that I was a teacher; it is also that I am an insider in terms of my relationships with teachers. I do not do research on them. My work with teachers often involves doing things with them (and sometimes for them) in their classrooms, facilitating professional development workshops for them (and sometimes with them), and when doing research I am most often a participant observer. I have taught a school mathematics class for a year and researched the classroom interactions; I have worked on curriculum resources with teachers and observed the implementation of those resources. I have co-researched with a classroom teacher as she enacted a particular emphasis she was interested in.

In the Observing Teachers project, I worked with three teachers, one of whom I knew prior to the study and had great respect for, and two of whom I immediately respected based on their experience in the classroom. When I first met the teachers, I learned that they agreed to participate in the study as a form of professional learning. Was it this combination of my history as a teacher, as a teacher educator, as a researcher, as a participant-observer, and as a person who understood the participants not as participants but as colleagues who had particular motivations to participate that resulted in the data collection looking as it did (and the themes that I am writing about)?

Lisa on Not Knowing the Participants

As I reflect upon my experience in this project, I struggle to articulate how different it felt for me with respect to the sort of research I usually do. My primary research focus is to address issues of equity, with a particular lens of the centering of Indigenous knowledges— ways of knowing, being, and doing—in mathematics education. I usually work within Indigenous research methodologies (IRMs) which require relationality with participants (Kovach, 2009). IRMs take seriously the importance of respect, relationship, reciprocity, relevance, and responsibility (Archibald, 2008; Kirkness & Barnhardt, 1991) and remind researchers that they must consider how their research gives back to the community (Kovach, 2009). My own research in Mi'kmaw communities in Nova Scotia has been guided by a community practice, *mawikinutimatimk* (coming together to learn together), and has emerged from decades long relationships rooted in trust and respect (Lunney Borden & Wagner, 2013). It is because of this relationality that I have confidence in doing the work that I do, and I also understand the importance of doing work that benefits the community, giving back.

In this project, my positioning was very different. I did not have a data collection site and, as such, I did not have a relationship with any of the teachers involved. Instead, I was given transcripts of teacher conversations and invited to make sense of their views on pedagogy from their utterances. I found myself wanting to know these teachers to better understand where their views were coming from and to try to connect with what experiences might have brought them to these understandings. There were moments when I was perplexed by a teacher's statement and wanted to ask for more clarity about the idea or value I saw emerging. There were times when I wished that I had seen what they were seeing as they viewed their videos to better understand the context. I found my position as an outsider in the research to be quite difficult relative to the work I normally do, and I wondered how much of my own views on education influenced how I interpreted the transcripts.

Furthermore, it felt unusual to interpret the data on my own. My normal practice, in using a decolonizing approach, is to invite participants into the process of analysis. We look at data together and try to interpret its meaning. We engage in a shared approach to making sense of data. Being alone with transcripts was a quite unfamiliar feeling. I invited a doctoral student, Ellen Carter, to join in the analysis with me for one of the chapters, which allowed us to engage in a modified version of this approach to data analysis. Being able to engage in conversation aided us in making meaning of the data, yet I still believe we both brought our views of pedagogy to the data. I have worried about how I represented these teachers, with whom I do not have a relationship. My hope is that I have done right by them.

Reflecting upon this experience, I wonder how it differed for my colleagues who were involved in data collection with teachers they knew. Did they interpret things differently due to knowing more about the context and/or participants? Did they see things in the data that I was unable to see because of my outsider status? Ultimately, I wonder about how research can be done well by outsiders and perhaps am increasingly more committed to the relational work that I do.

Comment

In these two accounts, Elaine and Lisa each reflect on aspects of the insider-outsider question. Elaine perceives herself mostly as an insider. She writes of her connections with the teachers, the schools, and the general milieu. She feels comfortable with these connections. Perceiving herself as an insider and perhaps being perceived as such is not simply a question of perspective. Elaine is clear about how this position is connected to her way of being during the research. When reviewing data from other regions, as well as noticing differences relating to teachers' pedagogy, she also noticed differences related to her way of conducting the focus groups as compared with those conducted by other members of the team. These distinctions lead to awareness about the various ways of being in the research setting and the different worlds that are brought forth as a result. For her, for example, the participants are colleagues, worthy of respect.

Lisa, meanwhile, perceives herself as an outsider. It happened that participant recruitment in some regions and languages was unsuccessful. This led to some team members, including Lisa, working with data collected by other members of the team. Lisa reflects on her uncertainty about this process, noticing a difference between this practice and her way of being and doing in her own earlier research. She describes her sense of missing the voices of the participants in conducting her analysis and interpretation of the data. Of course, as an outsider, it is still possible to do analysis, but the resulting findings must be understood as findings from an outsider perspective that reflect the structures and worldviews of the researcher. The knower and the known are always connected.

REFLECTIONS ON RELATIONSHIPS

A couple of themes run through the preceding reflections. The first concerns the insider-outsider question, which is apparent in all of the reflections, and we discussed it regularly during our meetings. The most obvious way in which insider and outsider perspectives are involved in the Observing Teachers project is in the generation of data by having teachers from within a region discuss videos made in their own classrooms. If there exists a regional pedagogy, then these discussions are of insiders discussing lessons made by other insiders. In the ethno-ethnography phase, when the teachers discussed lessons from another region, they did so as outsiders. In our research design, we positioned them intentionally as outsiders, and we were not concerned that their comments on the videos from the other regions would misrepresent teaching in those other regions, as our goal was not to analyze the lessons shown in the videos, but rather the pedagogy of the teachers making the observations. In this case, the teachers' outsider perspectives give us insight into their *own* pedagogy, just as reading ethnographies made by Europeans in the nineteenth century of natives of other lands gives us insight now into the thinking of nineteenth century Europeans.

An expected way in which insider and outsider perspectives are revealed by the project is in the analyses made by the researchers. We analyzed the data primarily as outsiders, but this was not our only perspective. Elaine, for example, describes her previous experience as a teacher in the same province and language as the teachers in her focus group; in her interactions with the focus groups, we can see her operating as a participant observer, taking an active part in the discussions. This is not so evident in the other researchers' interactions with their focus groups, but in our discussions as researchers we have observed that the researchers who were present for the focus group sessions feel more a part of *their* teachers' discussions than the other researchers who have only read transcripts of the sessions, as Lisa describes.

The second recurring theme is about relationships. All of the reflections address the nature of the relationships formed between research team members and teacher participants. The research design had less to say about these relationships in the mechanics of the anticipated data collection. Nevertheless, the enactivist perspective assumes that meaning is made jointly and contingently. The importance of these relationships is clear, for example, in Christine's and Elaine's reflections, which emphasize the importance of considering the teacher participants' learning through the research process, something Dominic also noticed. Relationships are also apparent in Annie's and Dominic's concerns for their participants' safety, while Lisa reflects on the unusual feeling of not having a direct relationship with the participants. Relationships are important in this project for a couple of reasons. The findings emerge from interactions (i.e., relationships) between groups of teachers and a researcher. In addition, these relationships have an ethical dimension. In all the reflections, we see concern for the teachers.

These relationships included relationships within the research team. As the project unfolded, for example, it became apparent that we each had different (sometimes very different) interpretations of our jointly authored research design. A good example of this point is the divergent ways in which the focus groups were conducted. We all agreed on the goals and guidelines for the focus groups in advance. In their enactment, however, they looked rather different. We must emphasize that we do not see this difference as a problem; we were not trying to produce cookie-cutter focus groups with standardized procedures (although Elaine's reference to a "violation" suggests that we may have thought otherwise). Differences are inevitable as each member of the team brings different histories and ways of being and doing to the project. Rather, then, our growing awareness of our different approaches led to discussions about the value of our data and the nature of our relationships with participants. Is it a problem, for example, that Elaine adopted a manner that she describes as an "active participant" while Annie's was more of an "expert"? Again, such differences are inevitable and, it should be emphasized, partly a result of our relationships with the participants. Elaine's role as an active participant was jointly produced with the teachers in her focus groups.

A further layer can be added to this last point. Our awareness of these differences arose from our observations of each other's research practice. As we each bring different sets of distinctions, so we see each other's practice a little differently (and so reveal a little of our own distinctions). Many of the discussions within the research team therefore revolved around clarifying these differences and distinctions, in order to better understand each other and ourselves.

What, then, is the value of research findings arising from a team of researchers who used somewhat divergent practices and interpreted data from a variety of individual frameworks? This issue has been explored in enactivist research. The notion of bringing forth a world of significance does not mean that "anything goes" and that we all see what we want to see. Rather, it suggests that knowing is related to the knower. Being aware of different knowers relating to the same object can display convergent features. Hence, researchers adopting different perspectives and bringing different histories may come to observe similar things. This happened in the Observing Teachers project, for example, when Christine and Richard conducted (with others) separate analyses of the same Ontario data (see Chapters 3 and 6). Their different histories led them to make different distinctions in the data (see Simmt & Kieren, 2015). When we compared the two analyses, however, we found that we had identified many of the same moments in the transcripts as being particularly significant. The resulting versions of middle school mathematics pedagogy included samenesses (student voicc) and differences (struggling with struggle).

CONCLUSION: ON THE TANGLED NATURE OF RESEARCH ON PEDAGOGY

Researching mathematics pedagogies is not straightforward. As a sort of culture, pedagogies are arguably impossible to observe in their entirety. They only exist as a distributed set of common practices across potentially hundreds of teachers, most of whom do not know each other. Yet commonalities can be observed. Pedagogy is difficult to characterize for an outsider because the observer may not see features that are important to local teachers. Pedagogy is also difficult to characterize for an insider because teachers may not see features that are distinctive when observed by outsiders. The Observing Teachers research design attempted to address this complexity by collecting data of teachers' observations of mathematics teaching in their region and in other regions of Canada. Our analyses of these teachers' observations are, however, inseparable from our own histories and perceptions as researchers. As such, our experience of conducting the project led us to reflect extensively on the insider-outsider question. More importantly, it led us to reflect and learn about the nature of the relationships we formed (or not) with participating teachers. We see these relationships as ethical in nature. Through the project we have, we hope, developed greater non-deliberate readiness-in-action (Brown & Coles, 2012) that we will carry forward into our future research.

REFERENCES

Archibald, J. (2008). *Indigenous storywork: Educating the heart, mind, body, and spirit.* Vancouver, Canada: UBC Press.

Berliner, D. C. (2002). Comment: Educational research: The hardest science of all. *Educational Researcher*, *31*(8), 18–20.

Brown, L., & Coles, A. (2012). Developing "deliberate analysis" for learning mathematics and for mathematics teacher education: How the enactive approach to cognition frames reflection. *Educational Studies in Mathematics*, *80*(1–2), 217–231.

Cochran-Smith, M., &. Lytle, S. L. (2009). *Inquiry as stance: Practitioner research for the next generation.* New York, NY: Teachers College Press.

Jaworski, B. (2003). Research practice into/influencing mathematics teaching and learning development: Towards a theoretical framework based on co-learning partnerships. *Educational Studies in Mathematics*, *54*(2–3), 249–282.

Kirkness, V. J., & Barnhardt, R. (1991). First Nations and higher education: The four R's—Respect, relevance, reciprocity, responsibility. *Journal of American Indian Education, 30*(3), 1–15.

Kovach, M. (2009). *Indigenous methodologies: Characteristics, conversations, and contexts.* Toronto, Canada: University of Toronto Press.

Lunney Borden, L., & Wagner, D. (2013). Naming method: "This is it, maybe, but you should talk to…." In R. Jorgensen, P. Sullivan, & P. Grootenboer (Eds.), *Pedagogies to enhance learning for Indigenous students* (pp. 105–122). Singapore: Springer.

Maturana, H. (1987). Everything is said by an observer. In W. I. Thompson (Ed.), *Gaia, A way of knowing: Political implications of the new biology* (pp. 65–82). Great Barrington, MA: Lindisfarne Press.

Maturana, H. R. (1988). Ontology of observing: The biological foundations of self-consciousness and the physical domain of existence. *Conference Workbook: Texts in cybernetic theory: An in-depth exploration of the thought of Humberto Maturana, William T. Powers, and Ernst von Glasersfeld.* American Society for Cybernetics Conference, Felton, CA, 18–23 October. Retrieved from: http://ada.evergreen.edu/~arunc/texts/cybernetics/oo/old/oo.pdf.

Reid, D., Brown, L., Coles, A., & Lozano, M.-D. (Eds.). (2015). Enactivist methodology in mathematics education research [Special issue]. *ZDM Mathematics Education, 47*(2).

Simmt, E., & Kieren, T. (2015). Three "moves" in enactivist research: A reflection. *ZDM Mathematics Education, 47*(2), 307–317.

Varela, F. J. (1999). *Ethical know-how: Action, wisdom, and cognition.* Stanford, CA: Stanford University Press.

CHAPTER 11

OBSERVING THE OBSERVERS

Reflections on Our Research

David A. Reid

In this chapter, I reflect on the origins of the Observing Teachers project. I make links to two facets of observer dependence: the dependence of the origins and conduct of the project on the researchers who originated and conducted it; and the methods we adopted and adapted to allow us to position the teachers as observers in the research. I reflect also on the object of the research, pedagogy, whose meaning is, again, observer dependent. We used one of those dependencies to gain insight into the teachers' meanings of 'pedagogy.' The dependency on our own meanings evolved from obstacle to advantage as we explored among our community of researchers our own backgrounds, assumptions, and values. Out of these reflections came some results and partial answers to our original questions, which are valuable, but not as valuable as the new questions we have come to ask.

INTRODUCTION

I spent most of my school years in Kentucky. At the start of Grade 6, I changed from the neighborhood elementary school to an alternative school in the inner city. The school had been founded a few years previously, under a director who had a vision of a school focused on self-directed learning, the liberal arts, and a close-knit learning community. She attracted a group of like-minded teachers to the school. This means that one of my first experiences of schooling, and the first

Researching Pedagogy and Practice with Canadian Mathematics Teachers, pages 231–239.

one I can remember clearly, was of teaching guided by a shared vision of teaching and learning. I would now call such a shared vision a *pedagogy*.

Another experience I had at the same time was of travelling from Kentucky to visit my relatives in Montreal, Canada and in England. I experienced the people I met in these three places as having not only distinct accents, but also distinct values and priorities. In my naiveté, I overgeneralized the differences I observed to being about all Kentuckians, all Canadians, and all English people. With more experience, I recognized that there is much diversity within these groups, but at the same time, I recognized further commonalities. English speaking Canadians I met in Montreal had read the same books as children as I had been raised in Kentucky. They were mostly not books other children in Kentucky had read, and not books that French speaking Canadians I met in Montreal had read. There were similarities in the differences, and differences in the similarities.

I became a mathematics teacher in Montreal, where I found the schools to be the same but different from schools in Kentucky. I then moved to Alberta for my doctoral studies. I engaged in school-based research, and I found the schools in Alberta to be the same but different from schools in Montreal. I became a professor of mathematics education in Newfoundland, where I found the schools to be the same but different from schools in Alberta. When the Observing Teachers project started, I was a professor of mathematics education in Nova Scotia, where I found the schools to be the same but different from schools in Newfoundland. During all these moves I continued, as I had as a child, to observe commonalities in the accents, values, and behaviors of people in the new places I moved to, that seemed different to me from the commonalities in accents, values, and behaviors of the people in the places I had moved from.

EVERYTHING SAID IS SAID BY AN OBSERVER

"Everything said is said by an observer"(Maturana & Varela, 1980, p. 8). I read that sentence for the first time when I lived in Alberta. It struck a chord. It seemed to me to pose a fundamental dilemma for research. What is the point of my sharing my observations of learning mathematics, if those observations are essentially about me as an observer, not about learning mathematics? I explored one way of addressing this dilemma with colleagues from Alberta, using the multiple contradictory but consensual perspectives of several researchers to say more than one thing about data we had all observed. After leaving Alberta, I never formed a similar community to pursue this approach (until the Observing Teachers project) and so I continued to think about other ways to do research, given that everything said is said by an observer.

After an embarrassingly long time considering this problem, the idea occurred to me that I should shift my attention from seeing observer dependence as an obstacle to research to seeing it as a feature of research. If everything said is said by an observer, could I not observe what others are observing as a way of learning something about those observers? More specifically, if I observed what teachers said about teaching I could learn about teachers. I imagined getting some teachers together, having them watch some videos of mathematics lessons and seeing what they said about them. I mentioned this (to me brilliant) insight to Christine Knipping, who, knowing more about comparative research than I did, suggested that Joseph Tobin had done something similar and I ought to look him up.

Tobin, it turned out, believed that there might be differences in what happens in pre-schools in Japan, China, and the US, and that these differences might have something to do with implicit values that were similar within each of those places, but different among them. He filmed events in pre-schools, showed the films to people from that place to see if it looked familiar to them, and then showed the films to people from the other places to see what they observed. His observers saw similarities and differences (see Tobin, 1999; Tobin, Wu, & Davidson, 1989).

While I was mulling over the issue of doing research, given that everything said is said by an observer, a great deal of attention was being paid to the results of international comparisons. In many Western countries, long-standing beliefs about the quality of education systems, or the inherent cleverness of that nation's children, were overturned, and many of the stakeholders in education were upset. In Canada, the results mostly revealed things that had already been revealed by previous national testing. Students in different provinces, and speaking different languages, performed differently. Reporters asked why, people speculated, and a number of researchers investigated factors that might matter, such as curriculum, gender, attitudes, beliefs, aspirations, time spent working outside school, parents' education, involvement, socio-economic status, and school resources (see, e.g., Anderson et al., 2006; Beaton & O'Dwyer, 2002; Schmidt et al., 2001; Wilkins, Zembylas, & Travers, 2002).

At roughly the same time that I had arrived at a way of doing comparative research in which observer dependence is a feature not an obstacle, an issue had arisen in mathematics education that called for comparative research. I contacted people I knew across Canada, we drafted a research proposal, and it was funded. That is when things became complicated.

PEDAGOGY

One challenge we faced in drafting the proposal was what to call the object of our research. Tobin had written about "'implicit cultural practices' of teachers ... practices that though not taught explicitly in schools of education or written down in textbooks reflect an implicit cultural logic" (Tobin, Hsueh, & Karasawa, 2009, p. 19). This was similar to Bruner's (1996) concept of folk pedagogy, the "taken-for-granted practices that emerge from embedded cultural beliefs about how children learn and how teachers should 'teach'" (p. 46). The teachers' observations of mathematics teaching would, we expected, give us insight into something implicit that guides their practice. But what was it? Tobin and Bruner both use the word *culture*, but that word carries a lot of baggage and we wanted something that was more focused on teaching practices, not everyday practices like holding doors open for strangers and apologizing when people step on your foot.

As Simmt notes in Chapter 2, the word *pedagogy* is not much used in the English language education literature, and especially not in mathematics education. The term pedagogy is connected to practice, but also has a theoretical connotation. We decided it would be a suitable name for the object of our research, if we were careful to explain the special meaning we were giving to it in our work. An example of an attempt to explain what we mean by pedagogy is the definition given in the introduction:

> Pedagogy refers to the set of (largely implicit) beliefs, knowledge, customs, and habits that guide a teacher's practice, and also the shared implicit beliefs, knowledge, customs, and habits of a community of teachers who interact regularly.

In the proposal, we also suggested that there was such a thing as *regional pedagogy* that we would seek to observe. The origins of my belief in the existence of such a thing can be found in the biography I sketched out. No one else involved in the project has the same biography, and so unsurprisingly, the degree to which the researchers believed in the object of our research at the outset varied.

We did have a theoretical basis for believing that groups of teachers who interact regularly would come to share *something*. Maturana's (1988) theoretical construct, emotional orientation, gave us another way to think about what we were studying. Of course, not every researcher was willing to take Maturana's word for it that there are emotional orientations that define domains of explanation, or that teaching is such a domain.

As our work progressed, the word *pedagogy* was much discussed; in Chapters 2 and 4, one can find different interpretations of the term. In Chapter 9, an alternative term, *professional vision*, is proposed that seems in many ways to fit the definition of pedagogy given above. Our clarification of the object of our research is a significant outcome of the project, though we have not arrived at agreement. Going back to the idea of multiple contradictory consensual perspectives, we have offered here a number of perspectives on pedagogy that we believe enrich the discussion, but do not finish it.

OBSERVING TEACHERS

I was very pleased with myself when I thought up the name for the project. I like ambiguous phrases that are correct in both their interpretations. The teachers participating would be participating as *observing* teachers. If *observing* is read as an adjective, it describes the participants and the special roles they would have in our research. What we would be doing was observing *teachers*. If *observing* is read as a verb, it describes what we would be doing as researchers. The problem with this sort of wordplay is that not everyone reads ambiguous phrases as having two meanings existing in some sort of un-collapsed quantum state. They pick a meaning and then note that an important aspect of the study (the meaning they did not pick) has been left out of the title.

METHODS

Tobin (1999) provides a thorough description of his research methods and we adapted them for our study. We briefly discussed developing a standardized set of prompts for the focus group discussions but did not develop this standardization for a number of reasons. First, standardization was inconsistent with the idea of multiple contradictory perspectives. If we all did the same thing, we might miss out on a question only one of us would think to ask. Second, agreeing on prompts would take time, perhaps a lot of time, and we did not need any further delays. Finally, it seemed unnecessary. The basic idea was to have the teachers talk about what they observed. We expected that they would not need much prompting to do that. This lack of standardized procedures led to some surprises, as we were more different in our perspectives than we knew. However, this also gave us insight into the observers that could have been hidden otherwise.

The methods of analysis show a similar diversity of perspectives. Again, this was intentional, as a fundamental principle of our methodology was to embrace multiple perspectives. For example, in Chapter 2, Simmt describes how she as a researcher guided and promoted the teachers' discussions, becoming herself a participant. This may reflect her personal connection to the teachers and her own background as a classroom teacher in Alberta. In Chapter 3, in contrast, Suurtamm and Quigley describe a much more 'hands off' approach in which "the conversations occurred naturally, rather than being prompted by researcher questions." Part II provides a clear example of three approaches to the same data and concludes with reflections on the differences among them and their observers. These kinds of differences are also discussed in some depth in Chapter 10.

The varied approaches to analysis adopted can be ascribed to individual backgrounds of the researchers, but it is interesting to speculate on the possibility that regional or linguistic research traditions could also be a part of the Canadian mathematics education research context. The same geographical and political factors used to motivate our hypothesis that regional pedagogies exist in Canada can also motivate a similar hypothesis concerning mathematics education research. However, we would expect the regionality of research approaches in Canada to be less strong than the regionality of pedagogies. Canadian researchers meet regularly at national and international conferences and, except in Ontario and Quebec, there are no research-oriented mathematics education conferences at the regional level. This is in contrast to the situation for teachers, who might meet at provincial or regional conferences, and rarely at the national or international levels.

There may be some regional variation in the theoretical perspectives adopted by researchers in Canada. Enactivism, for example, is most strongly associated with researchers in Alberta (Reid, 2014). Moreover, as is indicated in Chapter 4, theoretical approaches adopted from France have more currency in francophone universities than elsewhere in Canada. For example, in Chapter 4 the authors make use of the classic teacher-student-content didactic triangle (see, e.g., Chevallard, 1985/1991) in their analysis.

Canada can be contrasted with France and the Netherlands where there exist theoretical and methodological approaches that are widely shared at the national level. In France, Brousseau's Theory of Didactic Situations and Chevallard's (1985/1991) Anthropological Theory of the Didactic inform much of the research conducted there. In the Netherlands, Freudenthal's Realistic Mathematics Education plays a similar role (Sriraman & English, 2005). No such domestic theoretical perspective has been adopted in Canada. Instead, researchers in Canada freely adopt theoretical and methodological approaches from other parts of the world. Arguably, the only theoretical and methodological approach native to Canada is the enactivist approach employed in the Observing Teachers project, although significant contributions to developing this approach have also been made by researchers elsewhere (See Reid, Brown, Coles, & Lozano, 2015).

RESULTS

As noted in the introduction, large-scale assessments have revealed a wide range of student achievement in mathematics across Canada. Accounting for these differences has been the focus of research for many years, and factors such as students' attitudes, parents' education, and school curriculum and resources have been analyzed as possible influences on student achievement. The researchers suspected that pedagogy, because it directly influences teachers' practices, and hence presumably students' learning, should be as strong or a stronger influence compared to, for example, school curriculum and resources. Hence, we sought to address the question: How do pedagogies in middle school mathematics in regions of Canada differ?

However, the Observing Teachers research project suffered from the same basic weakness as studies that sought to account for variations in student achievement by examining the background information gathered in PISA on students, teachers, and schools, or information about curriculum and textbooks, or differences in government funding. Although pedagogy might be a strong influence on student achievement, so are many other factors, and so trying to identify the influence of any factor in isolation is fruitless.

Furthermore, our approach depended on an assumption that regional pedagogies exist. We ourselves have, at times, doubted this assumption. It is evident that the teachers involved in the focus groups each had their own perspectives. It may be that any commonalities shared by teachers in a region are hidden by the magnitude of individual differences among teachers. Nevertheless, there are some obvious commonalities among regions. The teachers had no difficulty recognizing the practices of the teachers from other regions as mathematics teaching. So perhaps our search for regional pedagogies was misguided because there appears to be a strong Canadian pedagogy, more important than any regional variation.

Alternatively, perhaps both individual teacher variation and Canadian commonalties are more significant than regional pedagogies.

In Part I several common themes emerged from analysis of the discussions of the focus groups. Engaging students, encouraging student participation, and taking a student-centered approach all touch on a focus on the learners in Canadian teachers' pedagogies. These themes are visible also in the comparisons in Part II. Teachers in all the regions also mentioned inclusion and embracing diversity in classrooms as important. This was further reflected in some of the comments made about using technology, that different modes of presentation might offer more students opportunities to engage with the lesson. These foci could indicate elements of a Canadian pedagogy for middle school mathematics.

While we acknowledge these points, we believe that the analyses in this book provide some evidence that the teachers, at least in the focus groups we observed, recognized that they have some things in common with each other, and that teachers from elsewhere seemed to think or act differently with respect to some aspects of mathematics teaching. This is reflected most strongly in Chapter 9, in which the anglophone teachers in Quebec explicitly discuss the differences they found in the videos they watched. It is also reflected in the comparisons made in the chapters in Part II. However, in many ways, these results are not the real outcomes of the project. The other issues raised and discussed in this book range more widely, and provide food for thought about the nature of teaching mathematics, of community, and of learning.

A JOURNEY, NOT A DESTINATION

I am writing this nine years after writing the original proposal for the Observing Teachers project, which proposed a three-year study. In the end, the funding was extended three times, and data collection occurred in places other than we had originally proposed. In some places, school districts were supportive of the project but teachers were not interested. In others, teachers wanted to participate but their school districts did not.

The project changed over time, and so did we, both individually and as a collective. Some researchers left the project and others joined it. For example, I was supposed to be the lead researcher in Nova Scotia but then moved to Germany. This created opportunities for others to take on new roles in the project and to add other researchers than those in the original proposal. Chapter 10 reveals not only something about us as observers, it also shows aspects of our work together in an emergent collaboration. We are now researchers with different perspectives, not only from each other but also from ourselves of nine years ago. We now know, collectively and individually, more about pedagogy, about possible regional differences in pedagogy, and about what mathematics teaching in middle school looks like in Canada, even if what we know is not what we expected to learn.

In 1996, I wrote about how researchers and the data they interpret co-emerge (Reid, 1996). Who the researchers are determines what data they observe, but at the same time, the data they observe changes who they are. Researching is a learning process, and like all learning processes, it changes the learner. In this book, we have reflected back on some of our learning and on the processes that led to that learning. Again, like all learning processes, this one has not ended. I am now a different researcher than I was when we proposed the Observing Teachers project, and I know that the research I do next will be informed by the Observing Teachers project and again change me.

REFERENCES

Anderson, J., Rogers, W. T., Klinger, D., Ungerleider, C., Glickman, V., & Anderson, B. (2006). Student and school correlates of mathematics achievement: Models of school performance based on pan-Canadian student assessment. *Canadian Journal of Education, 29*(3), 706–730.

Beaton, A. E., & O'Dwyer, L. M. (2002). Separating school, classroom, and student variances and their relationship to socio-economic status. In D. F. Robitaille & A. E. Beaton (Eds.), *Secondary analysis of the TIMSS data* (pp. 211–231). Dordrecht, the Netherlands: Springer.

Bruner, J. (1996). *The culture of education.* Cambridge, MA: Harvard University Press.

Chevallard, Y. (1985/1991). *La transposition didactique: du savoir savant au savoir enseigné* [*Didactic transposition: From learned knowledge to taught knowledge*]. Grenoble, France: Pensées sauvages.

Maturana, H. R. (1988). Reality: The search for objectivity or the quest for a compelling argument. *The Irish Journal of Psychology, 19*(1), 25–82.

Maturana, H. R., & Varela, F. J. (1980). *Autopoiesis and cognition: The realization of the living.* Dordrecht, the Netherlands: Reidel.

Reid, D. A. (1996). Enactivism as a methodology. In L. Puig & A. Gutiérrez (Eds.), *Proceedings of the Twentieth Annual Conference of the International Group for the Psychology of Mathematics Education*, Vol. 4. (pp. 203–210). Valencia, Spain: PME.

Reid, D. A. (2014). The coherence of enactivism and mathematics education research: A case study. *AVANT. The Journal of the Philosophical Interdisciplinary Vanguard, 5*(2), 137–172. DOI: 10.12849/50202014.0109.0007

Reid, D., Brown, L., Coles, A., & Lozano, M.-D. (Eds.). (2015). Enactivist methodology in mathematics education research [Special issue]. *ZDM: Mathematics Education, 47*(2).

Schmidt, W. H., McKnight, C. C., Houang, R. T., Wang, H. A., Wiley, D. E., Cogan, L. S., & Wolfe, R. G. (2001). *Why schools matter: A cross-national comparison of curriculum and learning.* San Francisco, CA: Jossey-Bass.

Sriraman, B., & English, L. D. (2005). Theories of mathematics education: A global survey of theoretical frameworks/trends in mathematics education research. *Zentralblatt für Didaktik der Mathematik (ZDM)*, 37(6), 450–456.

Tobin, J. (1999). Method and meaning in comparative classroom ethnography. In R. Alexander, P. Broadfoot, & D. Phillips (Eds.), *Learning from comparing: New directions in comparative educational research. Volume 1: Contexts, classrooms and outcomes* (pp. 113–134). Wallingford, United Kingdom: Symposium Books.

Tobin, J., Hsueh, Y., & Karasawa, M. (2009). *Preschool in three cultures revisited: China, Japan, and the United States.* Chicago, IL: University of Chicago Press.

Tobin, J. J., Wu, D. Y. H., & Davidson, D. H. (1989). *Preschool in three cultures: Japan, China, and the United States.* New Haven, CT: Yale University Press.

Wilkins, J. L. M., Zembylas, M., & Travers, K. J. (2002). Investigating correlates of mathematics and science literacy in the final year of secondary school. In D. F. Robitaille & A. E. Beaton (Eds.), *Secondary analysis of the TIMSS data* (pp. 291–316). Dordrecht, the Netherlands: Springer.

EDITOR AND AUTHOR BIOGRAPHIES

Claudia Anzovino, a McGill University graduate, is presently a Pedagogical Consultant for the Lester B. Pearson School Board in Montreal. After completing her education degree, she spent the next 19 years teaching at several elementary schools in Montreal. Claudia then went on to teach Mathematics and English at the secondary level for another 13 years with the Lester B. Pearson School Board and was also the high school's Mathematics Department Coordinator for the past 10 years.

Richard Barwell is Dean and Professor of Mathematics Education at the Faculty of Education, University of Ottawa, in Canada. Originally from the UK, he studied mathematics at the University of Warwick, before becoming a mathematics teacher. After a period teaching in Pakistan with Voluntary Service Overseas, he completed graduate degrees at the University of Bristol, where he Subsequently taught. He joined the University of Ottawa in 2006. His research foci include language, language diversity and discourse in mathematics classrooms. He also studies how mathematics education can contribute to a sustainable future for planet Earth. His work has been published in many journals and books. From 2010–2016 he was editor of the journal *For the Learning of Mathematics*.

Researching Pedagogy and Practice with Canadian Mathematics Teachers, pages 241–245.

Ellen Carter taught middle school and high school mathematics for ten years before returning to her own studies. She is now a PhD student at St. Francis Xavier University studying ways to support teachers of youth in care, while instructing courses in mathematics education and mental health.

Vandana Chandrasekhar is a PhD student and a Research Assistant at the Department of Education at McGill University, Canada. Her doctoral research investigates how university math methods courses support novice teachers to engage in appropriate teaching practices for student centered learning. Prior to pursuing her PhD, she had worked as a Math teacher at elementary and secondary schools in India, where she would often question and reflect on her own practice in engaging students in constructing mathematical knowledge. After taking an online course from Stanford University offered by Dr. Jo Boaler on Mathematical Mindsets, she was determined to seek ways of changing her own practice to support student learning which made her pursue her PhD.

Terry Wan Jung Lin received her Ph.D. in Education at McGill University in 2017. Her research interests are focused primarily on understanding and supporting professional learning in educational institutions. She has worked as a secondary teacher, an educational consultant, and a teacher educator. Currently, Dr. Lin is working within the Quebec Ministry of Education, in collaboration with the nine English-speaking school boards in the province, in building capacity for evidence-informed decision-making practices in the system.

Lisa Lunney Borden is a Professor of mathematics education at St. Francis Xavier University in Canada. Drawing upon her experience teaching in a Mi'kmaw community school for ten years, she is committed to research and outreach that focuses on decolonizing mathematics education through culturally based practices and experiences that are rooted in Indigenous languages and knowledge systems. She currently holds the John Jerome Paul Chair for Equity in Mathematics Education with a mandate to support mathematics learning for Mi'kmaw and African Nova Scotian learners. Her work has been published in numerous journals and books and she maintains the showmeyourmath.ca website.

Dominic Manuel is an Assistant Professor at the Campus Saint-Jean of the University of Alberta and a PhD Candidate at McGill University. He is originally from New Brunswick. He completed his baccalaureate in science with a major in mathematics, and in education in 2002 at the Université de Moncton. After teaching secondary mathematics, science and computer science for a few years, he underwent his masters studies and participated in multiple research projects. He also worked with local schools to support students who needed more challenges in mathematics by offering various enrichment activities. He obtained his master's degree in 2010 also at the Université de Moncton. He has also taught elementary and secondary mathematics education courses from 2006 to 2012 at the Université de Moncton and from 2012 to 2018 at McGill University as a course lecturer. He is currently finishing writing his PhD dissertation at McGill University. Dominic's research interest includes creating effective learning conditions by implementing inquiry-based learning in mathematics classrooms and using high-quality teaching practice during instructions. He is also interested in using inquiry-based instruction in mathematics and science in various ways including technologies to promote the development of mathematical and scientific talents of all students.

Scosha Merovitz works at Bishop's University in Sherbrooke, Quebec where she runs the Mathematics and Statistics Help Centre. In addition to teaching in the Mathematics Department, she teaches Mathematics Methods courses to future elementary and secondary school teachers. She is presently pursuing a PhD in Mathematics Education at McGill University. Her research interests include mathematics teaching communities; using instructional routines with pre-service and in-service teachers; using video as a tool for teaching, learning, and research; and embodied mathematics. Her most recent non-academic passion is tangoing the time away with her Greek husband.

Sophie Pinard completed her Masters in teaching and learning at McGill University, where she also worked as a research assistant. She is currently teaching high school mathematics at the Sir Wilfrid Laurier School Board in the province of Quebec.

Brenna Quigley recently completed her PhD at the University of Ottawa. Her SSHRC-funded doctoral research focused on literacy and assessment. Her dissertation presents a qualitative multi-method validity study that considers potential outcomes for critical literacy education in the context of a high-stakes standardized literacy test. She has also worked as a Research Assistant on numerous literacy and mathematics education research projects, including being a Project Manager for a study on mathematics teaching and learning. She has co-presented this research at many conferences. Since completing her studies, she has been teaching at the elementary and secondary levels.

David A. Reid is a professor of mathematics education at the University of Agder, Norway. His research interests include teaching and learning mathematical proof and proving, and methodological implications of the work of Maturana and Varela. Examples of his work include the book *Proof in Mathematics Education: Research, Learning and Teaching* (2010), co-authored with Christine Knipping, and a special issue of the journal *ZDM* on "Enactivist methodology in mathematics education research" (2015), co-edited with Laurinda Brown, Alf Coles, and Maria-Dolores Lozano. He edits the journal *For the Learning of Mathematics*.

Annie Savard is Associate Professor of mathematics education at McGill University. Prior to this position, she taught more than 15 years in elementary school. She is particularly interested in the contribution of mathematics in school to the development of citizenship competencies such as decision-making and critical thinking towards financial numeracy. She is interested in studying the development of probabilistic and statistical reasoning among young students and the use of technologies such as robotics to develop interdisciplinary knowledge. She also studies the professional development of teachers in regard to inquiry-based learning through Professional Learning Communities and the epistemological stances on pre-service teachers. She also works as an international expert in emerging countries. In 2019, she received a Professor Honoris Causa from Ovidius Universitate din Constanta in Romania for the accomplishments of her career in education.

Kathryn Sides Blanchard, a McGill Education graduate, started her teaching career in Montreal's Lester B. Pearson School Board. After teaching in elementary schools for a few years, Kate began working at John Rennie High School. At JRHS, Kate worked in the Resource and Math departments and found a passion working with students with learning challenges, particularly in Math. After nearly a decade at JRHS, Kate relocated to British Columbia with her family. Since leaving Quebec, Kate has worked in both the private and public sector. Kate is currently working as a Grade 7 Math and Science teacher in an innovative learning community at Rutland Middle School in Kelowna, BC.

Elaine Simmt a Professor of Mathematics Education and Associate Dean at the Faculty of Education, University of Alberta. She is an Albertan, and except for a very short period of time when she lived in British Columbia she has lived in Alberta her whole life. Elaine is a former schoolteacher of mathematics, chemistry, and physics and has been working with teachers and curriculum leaders for some 25 years. Although her own education and teaching experience has been located in Alberta she was the director of two international teacher development projects. Since 2010, she has worked in Tanzania on projects that ask how universities can work with communities (professional and civic) to build capacity for mathematics teaching in rural communities. In the last 4 years, she has been working with Omani teacher educators to develop teacher training modules for in-service teachers. That work has become a site for researching international collaborative work. Working internationally has provided Elaine with an opportunity to reflect on curriculum and pedagogy as grounded in the local context. It is not surprising that her work in this project took a turn to a more theoretical question about the nature of the notion of pedagogy itself.

Christine Suurtamm is the Vice-Dean of Research and a Professor of Mathematics Education at the Faculty of Education, University of Ottawa. She teaches in the areas of mathematics education, assessment, and qualitative research. Her research focuses on the complexity of mathematics teachers' classroom practice, with particular interest in teachers' formative assessment practices as opportunities to attend to students' mathematical thinking. She is the Director of the Pi Lab, a research facility funded by the Canada Foundation for Innovation. She has been the Principal Investigator on several large-scale projects focusing on mathematics teaching and learning, was the Canadian representative on the board of the National Council of Teachers of Mathematics (NCTM), and was Co-Chair of Topic Study Groups on Assessment at the past two International Congresses for Mathematics Education (ICME-12 & ICME-13). She has won several awards for research and teaching. She is a co-editor of the series *Research in Mathematics Education*, published by Information Age Publishing.

Printed and bound by CPI Group (UK) Ltd, Croydon, CR0 4YY

15/09/2026

14962336-0002